Designing
with video

Salu Ylirisku & Jacob Buur

Designing
with video

Focusing the user-centred design process

 Springer

Salu Ylirisku, MSc
School of Design
University of Art
  and Design Helsinki
Hämeentie 135 c
00560 Helsinki
Finland

Jacob Buur, PhD
Mads Clausen Institute
  for Product Innovation
University of Southern Denmark
Grundtvigs Allé 150
6400 Sønderborg
Denmark

Proofreading by Cindy Kohtala.

Layout design by Kalle Järvenpää.

British Library Cataloguing in Publication Data
Ylirisku, Salu
   Designing with video : focusing the user-centred design process
   1. Engineering design 2. Video recording 3. Computer-aided design
   I. Title II. Buur, Jacob
   620'.0042'0284
  ISBN 9781846289606

Additional material to this book can be downloaded from http:/extras.springer.com

ISBN 978-1-4471-6850-8   ISBN 978-1-84628-961-3 (eBook)
DOI: 10.1007/978-1-84628-961-3

Printed on acid-free paper

9 8 7 6 5 4 3 2 1

Springer Science+Business Media
springer.com

Preface

Sometimes lack of knowledge can be a blessing. If I, Salu, had known how much it requires to piece together this book, a certain desperation might have overwhelmed me at the moment of beginning the to write. However, the great effort this book has taken stems largely from the dialogical character of authoring. The process was essentially a long and intensive text-mediated and video-mediated learning activity. Yet, one day the work must be said to be completed. And here it is.

The book would never have reached this stage without the long experience of Jacob Buur, who has conducted numerous exciting research projects where new uses for video have been explored. Moreover, his wisdom in guiding the co-authoring of this book with me has provided a most wonderful opportunity for learning about writing, about video-making and about the philosophy of user-centred design.

This writing process has provided an extraordinary channel for learning. I feel that only now do I know how little I know. I am indebted to the work by the early pioneers, such as Wendy Mackay, Brenda Laurel, Brigitte Jordan and Austin Henderson. Without their willingness to share their ideas and expertise, the book would be of much less value.

This book is about video and its relationship to design. Especially, the focus is on how it can be taken as a tool for driving design. Video has a remarkable impact on people. However, where this impact actually resides is not so obvious as the following story shows.

When this book was almost finished, I had a colleague visiting one morning at quarter past seven to capture a short video for a workshop with a national children's association, on the communication between parents and children. We had agreed on the time of the visit two days previously. Before the event I suddenly recalled the comment by Austin Henderson during a dinner with Jacob Buur. He had participated in a study that focused on interaction with a new collaborative tool, and he said that if he had known his actions would be analysed in so much detail, he perhaps would have declined to participate in that study. The video was so revealing.

In the evening before the videotaping I had already begun to mull over the morning. Would I take the usual time to read the newspaper? What would I wear? Would I quickly check my e-mails though it might make me look "selfish" on the video? Am I the kind of parent that just reads papers and checks e-mails despite having two gorgeous children to take care of – and this is why the person with the video camera is there?

When my colleague arrived, we were already all awake, and my wife was just leaving for work. "Pasi is here!" my children shouted. He is quite familiar to the children, and his arrival transformed the morning into something quite different from the usual one. When the camera was put on *rec* and the shooting began, my five-year-old daughter started the show about how well she does everything by herself. The younger one (one and a half years old) was continuously looking at what the cameraman was doing. We had not had that easy a morning so far as regards putting clothes on. I, however, forgot to brush both the girls' teeth.

So, perhaps more than the audience, video influences its creators – all the people who are present in the making of the video artefact. In this book we try to outline a more conscious user-centred design practice that is sensitive to how people collaboratively learn and become inspired by the user's reality, and how the authoring, moulding and sharing of video artefacts help to achieve the desirable changes that designers are after. We also aim to illustrate how video influences the user-centred practice through a rich variety of cases, method descriptions and some bits of theory.

The book was authored in collaboration with numerous case authors, whose contribution was invaluable. They provided concrete examples and helped us learn more about how video may be employed in design. Thank you for this! Special thanks to Turkka Keinonen, who initiated the idea for this book. Antti Raike, who created the cinemasense website (at *http://eloku-vantaju.uiah.fi*), has greatly inspired a closer look at what moviemakers and movie theory may offer design. Thanks also to Tuuli Mattelmäki, a long established colleague, who one day asked me to join the design research group at the University of Art and Design Helsinki. Many thanks also to those who participated in the various workshops that we arranged around this book and who helped to grasp the essence of the role of video in user-centred design. Thanks are due to TEKES (The Finnish Funding Agency for Technology and Innovation) for funding the related work at the University of Art and Design Helsinki. Finally, many thanks to those of you (users, designers, managers, and others) who have let yourself appear on the videotapes!

To return to where I started, the lack of knowledge: ignorance may help one to avoid being nervous, but it certainly does not help in professional design to construct something good for people.

Salu Ylirisku,
Helsinki in January 2007

Contents

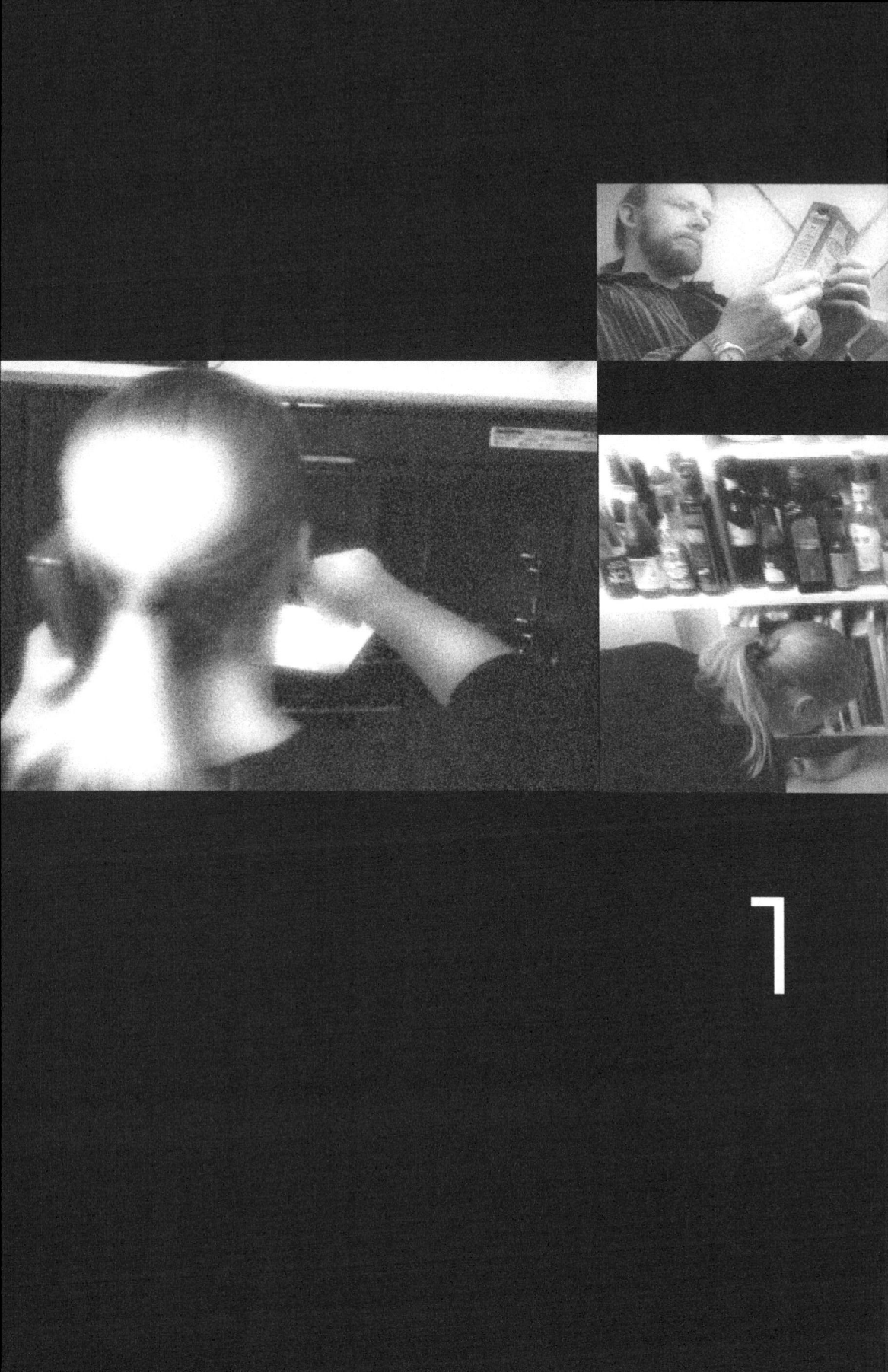

1

Video
in design

YRJÖ ENGESTRÖM

*"Object-oriented actions are always,
explicitly or implicitly, characterized
by ambiguity, surprise, interpretation,
sense making, and potential for change."*

1

Video
in design

"The creative treatment of actuality" – this is how John Grierson, the father of the British documentary movement, characterised documentary filmmaking already in the 1920s. His insights were developed some 30 years after the moving image was first invented by Thomas Edison. The statement reflects a deep understanding of the maturing relationship between people, reality and the moving image. The earliest documentary films date back to as far as the 1890s[†], when the Lumière brothers captured short scenes with their newly invented cinématographe. Their first films showed brief scenes from everyday life such as the famous "Workers Leaving the Lumière Factory", which displays a crowd of people walking through the factory corridor. These films were unedited moments from real life – *actualities* – as the Lumières called them.

These scenes, however, lost their attraction soon after the first wave of excitement, and the moviemakers had to explore new ways to regain audience interest. Techniques such as fictional narration, staging, continuity editing, montage, and camera movement developed during the early 20th century. Movie technology has since evolved from the large theatre screens through to television receivers into the tiny displays of mobile devices. With the advent of the computer, cinema grew from optically reproduced images of reality into computer-based constructions of virtual realities. Movies are still created in all these forms, and the variety keeps expanding. The realm is huge, but what is its value for product design?

[†] The question of who captured and showed the very first film, and where, is a matter of controversy, since there were many simultaneous efforts in process on two continents during the 1890s.

This book is an attempt to answer this question. The underlining principle throughout the book is interactivity. Cinema, including video, film and TV, are of little value if perceived as a monologue. In such a form people are perceived as passive recipients of the data flow on a screen. Glued to their chairs they merely *look* at the moving picture. But looking is not seeing! "Looking is a biological act: Open eyes look. Seeing is an act of conscience," wrote the pioneering theatre director Augusto Boal (1998, p. 79). Seeing is about consciousness and interpretation. And *this* is something that can be fostered in an interactive design process with video. When video is adopted as a tool to facilitate a more conscious design process, it truly turns into a mediator of the "creative treatment of actuality".

User-centred design in transition

User-centred design is an approach to designing products, systems or services that puts the people who will use the product at the centre of the development effort. The approach promotes the active involvement of potential users of the designed products in the process. The overall goal of user-centred design is to ensure that a product has potential in the market and that it improves the quality of life and work as perceived by its users. Henry Dreyfuss, one of the first industrial designers in the US during the early 1900s, crystallised the aspirations of user-centred designers in the following words (Dreyfuss, 1967):

> ...if people are made safer, more comfortable, more eager to purchase,
> more efficient – or just plain happier – by contact with the product, then
> the designer has succeeded.

The quote reveals that user-centred design is not an especially new invention. The term "user-centred design" was coined in the late 20th century to elevate the awareness that the influence modern computerised and industrial products had on their users. The first international conference with an essentially user-centred focus was organised already in the early 1970s (Cross, 1972). The topics covered such areas as "social technology", "participation in planning", "adaptable environments" and "computer aids". The user-centred design field has since grown into a global business backed by international networks of academic research. The main topics, however, have essentially remained very similar but with a substantial growth in detail and depth.

Along with the spread of user-centred design, a major shift has begun to take place in design thinking. We are moving from perceiving design as a problem-solving activity to understanding it as the social construction of new opportunities. This change has bearings on how the role of users – or everyday people – is being perceived in the design process. Users are increasingly becoming key collaborators to drive innovation and strategic decision-making in industries (Keinonen and Takala, 2006). Rather than being "involved helpers" in a collaborative problem-solving business, users are turning into co-developers in the design process.

Modern interactive products, such as flight reservation systems, have become extremely complex. They require an understanding of both the various people involved in the flow of activities, as well as the related technical systems. At the same time, work has become increasingly specialised. These developments present accumulating challenges in the process of building an understanding of design: What do people do, what do they value, and what do they want? How will the solutions fit into practice, and how will the people adapt their behaviour and react to the new technologies? These challenges demonstrate the argument that new collaborative methods are needed to perceive the intended change against the backdrop of current reality.

Numerous methods for this purpose already exist. These methods, however, often focus only on part of the whole: the study of the existing use context, the participation of users, or the empathic and experiential understanding of how the world is perceived by users. Video is a tool that helps to bring these aspirations together. The ways in which video can do this is outlined in the following pages.

Good products – a proper aim?

User-centred design aims to deliver good products to markets. Creating a product that is too difficult to use, or that does not serve the aims of people, can have a tremendously negative impact on business – both the manufacturer's and the users'. A new product design project is usually a significant investment for a company, and the user-centred approach is one way to minimise the risk of launching a "wrong" product. From the users' point of view poorly designed products make life more difficult and irritating and work less efficient. People have long memories. Once dissatisfied with a product's performance, it is likely that they will choose a competitor the next time they purchase the item.

What makes a good product? And what contributes to the creation of a good product? Despite the attraction of the topic, we will here not delve deep into the debate of what good products are. Instead we shall explore the underlying issues. Issues such as the fundamentals of *design* and *evaluation* may be more fruitful to understanding the role of video in product design – especially in the fuzzy front-end of deciding which product to create.

"Design" is a term loaded with meaning. John Heskett, an internationally recognised scholar of design history, elaborated the ambiguity of the term "design" thus (Heskett, 2002, p. 5):

> *Design is to design a design to produce a design.*

The word "design" is conceived as a noun with three different meanings: the field of design, a conceptual proposal, and a concrete product. It is also seen as a verb to denote the activity. Three of these meanings are especially interesting here. *First*, we use the term "design" to refer to the *result* of design work, *i.e.* to the *change* that a design project creates. A change can be many things: An improvement to an existing solution, such as the airbag in a car, or the roller in a computer mouse; a change in the physical appearance of a product; a change in how the thing works or is operated. Creating something completely new – whether a physical object, a service, or a system – is also to make a change. *Second*, this change is usually conceptualised before it is actually realised. The design is the conceptual idea, or theory, about what may be valuable for people. *Third*, "design" refers to *an activity* or *process* that aims at creating these changes.

Designs as *changes* comprise a broader perspective than when seen as *objects*. In architecture and industrial design there has traditionally been a very strong focus on the artefact itself: design prizes are given to objects, and the designs are exhibited "pure", away from the messy context of daily use or – in the case of architecture – without people obscuring the beauty of the artefact. This way of perceiving products is desperately narrow, since it completely ignores how products will function in situ. Introducing a new design changes its environment and transforms the practices of the people that face it. For example, the introduction of the digital camera has changed how we take and share photographs. Understanding products as changes shifts the focus to exploring what these changes are and how they appear in the context of use.

Speaking in terms of change, we shall define *evaluation* as the process of perceiving the character of change. Evaluation distils and verbalises the merit, worth, and significance of the change. In design literature evaluation is usually focused around the established "system specification" or "product requirements", and the formation of the evaluation criteria in the early steps of the process are explicitly emphasised. However, when the design organisation is unsure about which product should be created – *i.e.* in the fuzzy front-end of product design – the formal specifications and requirements tend to provide a too rigid framework and vocabulary for understanding the process. The process of negotiating the evaluation criteria, fundamental to perceiving how good a product is, is often omitted, or is quickly passed over as something too difficult to describe.

It is thereby the negotiation of the evaluation criteria, rather than the evaluation of a product according to some criteria, that takes place in the early design phases. The process of negotiating requirements and specifications, and the formation of early product ideas and visions, are the specific areas of interest in this book. This area lies notably beyond mere requirements and specifications and is fundamentally a social and political endeavour. The criteria negotiation is coloured by the values of the organisations and people involved, and entails making choices over other people's capabilities and constraints: Would someone in a wheelchair be able to use this product? Would the users be able to use the product with one hand? What skills are required for using it? How much space does it take? The particular point of view of user-centred design is to perceive the products in relation to the *use contexts* and construct the evaluation criteria with emphasis on the impact that the products have on the users' lives.

How many users are needed, what environments should be studied, and who should be included in any study on building appropriate evaluation criteria for a product? This is a question of deciding on the *relevant use context* that needs to be accessed by the design process. The next question is about the relative importance of all the things that are encountered in the process. For example, how important is the clothing of the users? How much does it matter that they are already carrying a mobile phone? Are the users' informal meetings in the corridor important to the design? These are all questions that may be faced by a design team entering the users' reality in order to inform their design process.

Throughout this book we shall continuously encounter the term *use context*. What does it mean? The Collins COBUILD dictionary (1987) defines "context" in the following way:

> *The context of something consists of the ideas, situation, events, or information that relate to it and make it possible to understand it fully. If*

something is seen in context or if it is put into context, it is considered
with all the factors that are related to it rather than just being consid-
ered on its own, so that it can be properly understood.

The above broad definition gives little practical hint as to where to focus in a design project. What the definition does highlight, however, is that designers need to not only perceive a product but to see the diverse ways it relates to the texture of the everyday life surrounding it. When a product idea is not known in advance but is being constructed during the project, the framing of the context is a matter of exploring and discussing what there is now and how important it is. The context may trigger new product ideas and influence those that have already been crafted. Thus, the context also affects products. According to a more radical interpretation, everything that a design affects, or that the design is affected by, forms *the relevant context* of design.

From a practical point of view, the key ability of designers and design processes becomes the skill to foresee the entire situation that arises when a new product is introduced to a social setting: What will change? How will people adapt their daily activity? Confronting this challenge requires new tools that are able to bring the use contexts into design in more varied ways. This is a particularly suitable role for video to play. It is not however enough to merely bring in more detailed material and greater amounts of it, but to seek different perspectives and see how the pieces of the puzzle may relate to each other in novel ways. This dialogue between designers and the materials of the situation is fundamental to designing (Schön, 1983). The dialogue is becoming more and more social as the amount of information in design projects has long ago exceeded the capacity of individual designers. As environments become increasingly populated by smart and networked devices, social practices evolve in every area of life, and skills and expertise are dispersed throughout organisations; a social process is necessary. And video is essentially a social tool.

Does video solve design problems?

The traditional approach to design is to understand it as an activity aiming at solving *design problems*. Design projects are launched to overcome a problem with current solutions. The problem may be a technical fault in the product, or it also might be an economic or ecological problem, or a problem with

production. Bruce Archer, the former director of research at the Royal College of Art London, stated in 1965 that the proper way to proceed with design
is to first express the design problem in terms of *abstract criteria* together
with listing *all the factors* affecting the design, and then divide the design
problem into sub-problems and solve these in a prioritised order (Archer,
1965, in Cross 1984). A clear definition of the problem focuses the activity
to solve it, and solving clearly stated problems appears efficient. This is perhaps the main reason why the problem-oriented approach has dominated
design thinking for decades, and still does.

Understanding design in this way pre-supposes that a design problem,
and all the factors affecting it, *can* actually be identified before solving it.
However, this is not how design happens in real projects, and in the light
of contemporary thinking it seems that this is not even possible. Design
problems are married to their solutions. This idea was present in Archer's
(1965) thoughts but was overrun by positivistic thinking, which embraced
the partitioning of tasks for efficiency and control. The marriage of problems and solutions became widely acknowledged when Rittel and Webber
(1973, in Cross 1984) published their study on the "wicked" nature of design
problems. Planning problems, such as those in design, are ill-structured, or
wicked, rather than being closed problems with a single solution. According
to Rittel and Webber (1973, in Cross 1984, p. 137):

> *The formulation of a wicked problem is the problem! The process of for-
> mulating the problem and of conceiving a solution (or re-solution) are
> identical, since every specification of the problem is a specification of the
> direction in which a treatment is considered.*

This statement renders antique the terms "problem" and "solution" as central characterisations of the essence of design. We can then start to talk about
design as a *focusing activity*. Focusing is the activity of clarifying a design
challenge. Focusing is thus a *goal-directed activity*. During the process focus
transforms from being broad and blurred towards a sharper picture of the
relevant issues. The aim of focusing is to discover the *valuable and meaningful* issues for the people and organisations involved.

This step towards understanding design as a focusing activity is also a
step towards the etymological history of the word "design". According to
Krippendorff (1996) the word "design" has Latin origins. It is an amalga-
mation of the words "de" and "signare", the combination meaning "making

something, distinguishing it by a sign, giving it significance, designating its relation to other things, owners, users, or gods" (Krippendorff 1996, p. 156). This suggests design as a process of making sense of things.

Video provides a tool to collaboratively build conceptions of (*i.e.* conceive) design opportunities while keeping our feet on the ground of reality. The term "conceive" also derives from Latin, from the word concipere: "to take in

From the On- †
line Etymolo-
gy Dictionary
at http://www.
etymonline.
com/

and hold". One of its original meanings is also "to take (seed) into the womb, become pregnant".† Both conceiving and making sense are essential to creating new ideas. These activities are also fundamental to understanding how the new ideas will influence their surroundings and eventually the reality of people. A process that is built on a dialogue with these activities becomes a more *conscious process*. Such a process enables the true making of choices among many alternatives. Without consciousness design is blind.

Guided by surprise

Surprise is a wonderful indicator of the potential for new learning. Designers attempting to develop a more conscious design practice thus benefit from understanding how surprise helps identify and modify presuppositions and expectations. Psychologist Jerome Bruner (1986) stated that surprise is something that necessitates the existence of a coherent structure for expectations: "...surprise is a response to violated presupposition" (Bruner, 1986, p. 46). Expectations guide our actions rather automatically, even subconsciously. It is therefore highly beneficial to understand our own biases and to consciously develop greater skills in sensitive reflection.

The skill to synthesize, to discover and establish coherent structures in the world is the other side of the coin to experiencing a surprise. Such structures include what the psychologists call the "expectations". According to Bruner (1986, p. 48): "The virtue of such models is that they enable us to keep an enormous amount in mind while paying attention to a minimum of detail." However, the larger and the more rigid the structure becomes, the less responsive it is to fluctuations in individual situations. The costs to change the structure increase as it matures. Thus, a design team needs to explore the alternatives and test their models at an early phase. The following example underlines the importance of this.

"Expectation" is a rather close relative of "focus". Expectations and focus both guide exploration. Exploration reveals new issues, which are then related to expectations. As a result, the expectation may be modified: changed or enforced. Donald Schön (1983) studied designers' activities with the focus

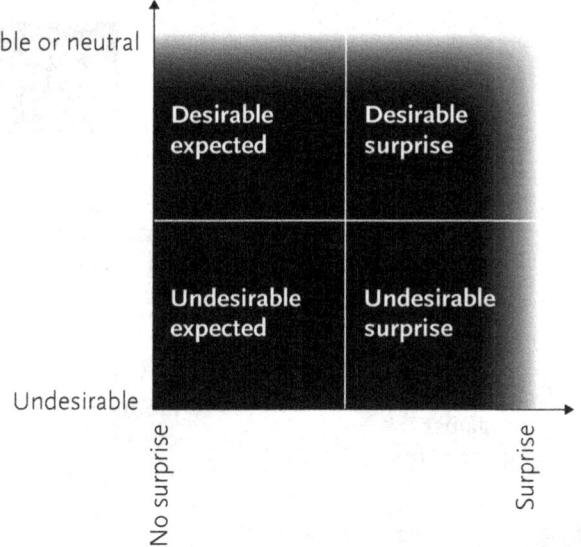

Desirable or neutral

Desirable expected

Desirable surprise

Undesirable expected

Undesirable surprise

Undesirable

No surprise

Surprise

1 Video
in design

on "design reflection". He outlined the following picture (Figure 1.1) of the relationship between design moves and observed outcomes.

A designer's skill to reflect upon a situation improves when more situations are encountered and considered. Schön (1983, pp 140) states that

The artistry of a practitioner ... hinges on the range and variety of the repertoire that he brings to the unfamiliar situations. Because he is able to see these as elements of his repertoire, he is able to make sense of their uniqueness and need not reduce them to instances of standard categories.

Perceiving a situation against the experience of another earlier situation enables a designer to compare the *similarities* across the situations as well as the *differences* between them. Seeing one situation as another is not enough. It offers a practitioner a new way of seeing the situation, but the appropriateness and value of this new perspective needs to be evaluated. Schön asserts that this can only be truly done by experimenting. Reframing the design challenge by seeing it as another recasts the form and the relationships of the design opportunity anew. It gives new resources to the design team in evaluating the move against five questions that Schön outlines (pp. 133 and 141):

▸ *Does a reframing help to approach a coherent solution?*
▸ *Does the design team value the result that it helps to achieve?*

Figure 1.1
The possible consequences of a design move in relation to intentions (adapted from Schön, 1983)

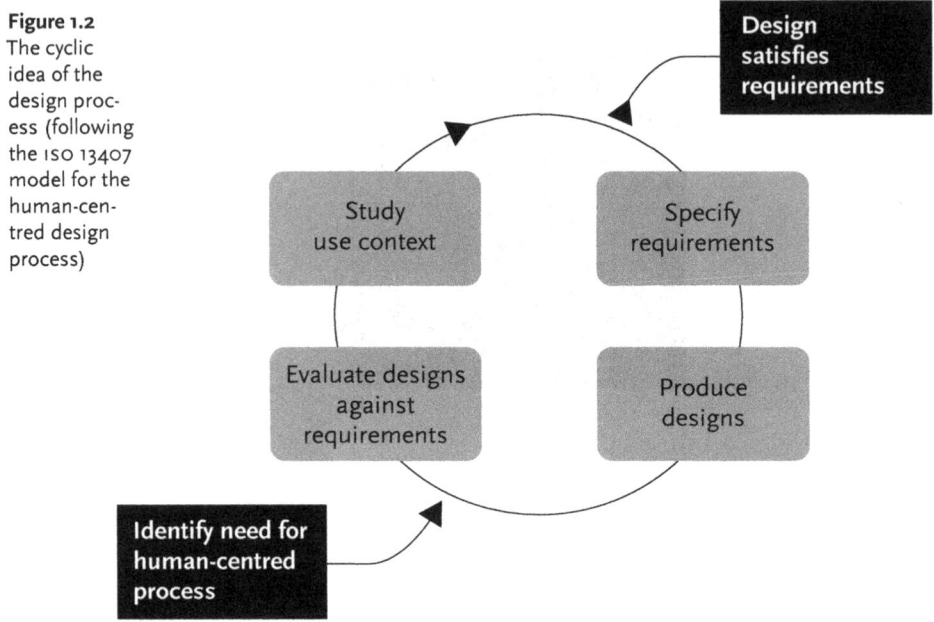

Figure 1.2
The cyclic idea of the design process (following the ISO 13407 model for the human-centred design process)

Design satisfies requirements

Study use context

Specify requirements

Evaluate designs against requirements

Produce designs

Identify need for human-centred process

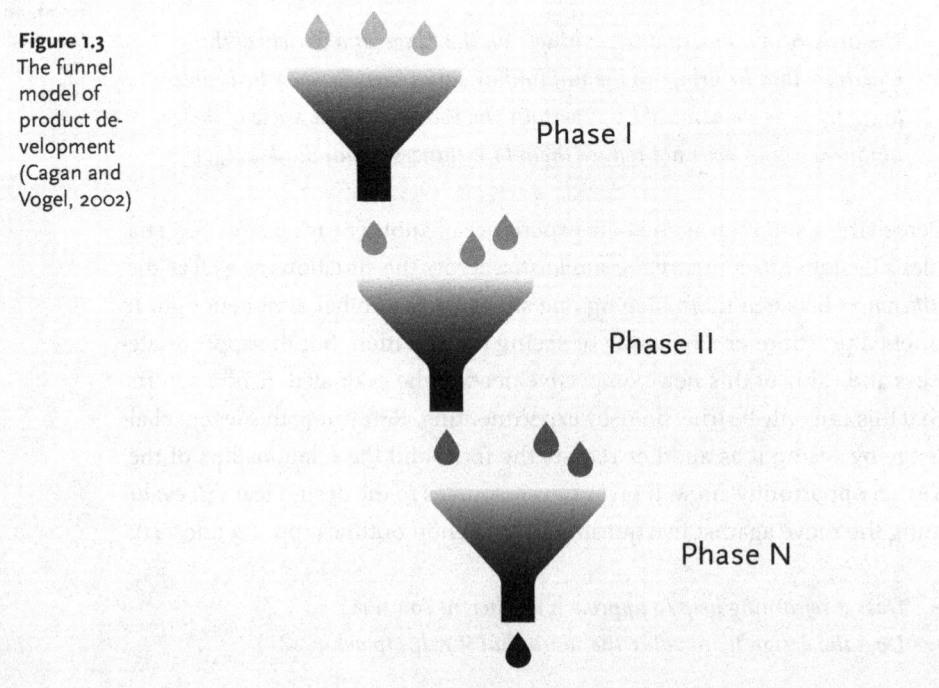

Figure 1.3
The funnel model of product development (Cagan and Vogel, 2002)

Phase I

Phase II

Phase N

- ▶ *Does it guide towards a more coherent idea about what the design challenge is?*
- ▶ *Is the result congruent with the fundamental values and theories of the design team?*
- ▶ *Does the result have the capacity to keep the inquiry moving?*

As the ability to reflect develops with experience and is highly domain-dependent, the design process benefits greatly from the input of experienced designers as well as the experience of the other people that participate in the process. For example, experienced workers (in the case of a worker-themed user study) may have a broad repertoire of possible situations to bring to the design events. These issues underline the need for the *collaborative* design practice that will be addressed in subsequent chapters.

Where does video fit?

It is generally assumed that professional design proceeds in a systematic way. Different design projects actually run with very dissimilar processes as the structure of design activities depends heavily on the design context. Design literature is saturated with design process models, which seek to better structure design, and thereafter, make it more controllable, predictable and efficient. Some process models such as the one presented by Ulrich and Eppinger (2003) describe the process as a rigorous hierarchy of activities with clearly stated phases following each other in a particular order. The ISO 13407 (1999) Standard Human-centred Process for Designing Interactive Systems differs in that it presents a small set of activities in a cycle, where the activities follow each other iteratively. The cycle contains phases of understanding and specifying the use context, specifying the user and organisational requirements, producing design solutions, and evaluating designs against the requirements (ISO, 1999) (see Figure 1.2).

Cagan and Vogel (2002) describe a rather similar model, an "integrated new product development process", which also comprises four phases: identifying, understanding, conceptualising and realising product opportunity. These phases are presented as funnels that each receive multiple inputs and produce a single result for the following funnel (Cagan and Vogel, 2002). Common to these models is that each is explicitly directed to manufacturing a product, and they are comprised of steps that follow each other in a particular order (see Figure 1.3).

Product development process models stemming from the engineering design tradition such as the ones above are likely to be too rigid for projects that aim at radically new solutions, affecting market expectations, enhancing decision-making, or at profound organisational learning (Keinonen and Takala, 2006). The engineering design models impose a strict structure on the order that the design activities follow. The projects in this book however are all basically early design projects, most focusing on conceptual design. In such projects, the activities unfold in an arbitrary order compared with the process models, and moreover, the activities take place in parallel. For example, the "Freeride skiers" case study in Chapter 2 began with the idea for a prototype, and then continued with a user study, while the development of a functional prototype advanced in parallel. It then proceeded into further user studies with a very open focus to produce background knowledge for several product generations. These kinds of projects cannot be managed with conventional stage gate control because the structure of activities will be different in each case and people's roles in the process vary dramatically. Hence, placing video into the design process is rather difficult.

Rather than depicting a model with definitive phases, a dynamic framework is presented. It accommodates the main activities in a user-centred design process while allowing the flexibility of real projects. The process proceeds from the past, from the historical body of reality, towards the future, *i.e.* realising the change that a project strives to achieve. The intersection where the line entwines forms four spaces. These spaces present the activities related to focusing: *exploring, describing* and *relating*. At the centre is *focus*. Exploring encompasses the discovery into past, present, and future as displayed in practices, materials and spaces, and people's thoughts about these. Relating refers to the activity of connecting the discoveries explored to the other materials that are already known, and studying the relationships of the emerging groups of themes. Creating is the activity of forming new concepts and ideas, and combining these into concrete new structures.

The activities of exploring, relating and describing are intrinsically intertwined. The model, as depicted in Figure 1.4, is a broad, general one, and it scales from individual events to large projects. A design process described at this level strongly resembles the reflective thinking outlined by John Dewey (1910). Thus, the process can also be called the grounded co-thinking process.

This book is organised with the above structure in mind. Chapter 1 explains the current state of research. It also provides a review of known prac-

Figure 1.4
A framework
of the aspects
of designing

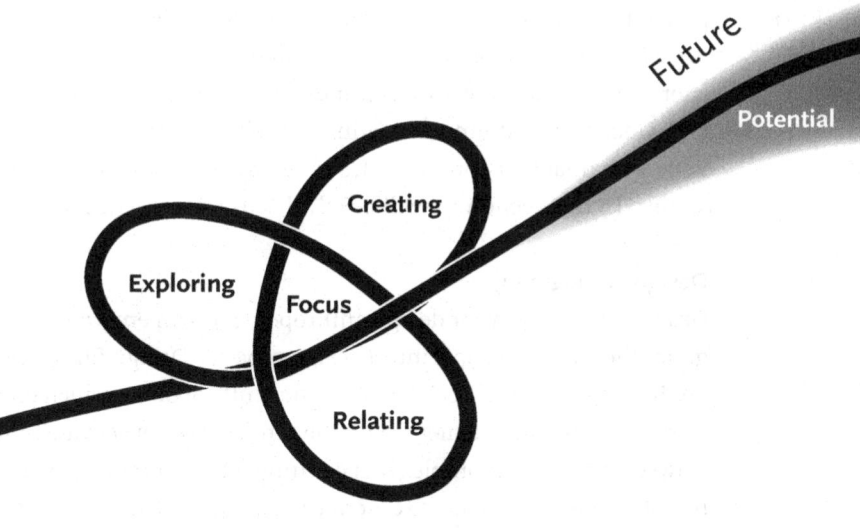

tices concerning the use of video in earlier projects. The second chapter details how video is employed to explore the reality of users. Chapter 3 describes processes for relating materials and ideas. The fourth chapter moves towards describing ideas on potential products. The last chapter presents how video provokes people to realise the opportunities described.

The activities may begin with a user study, proceed with the interpretation of the findings and clarifying the design challenges, create concrete visions and proposals, and then implement these visions. These phases may occur in any other order as well. Hence, a normative structure concerning the "right" way to proceed in a design process will not be proposed here.

Video traditions in design

When discussing how to utilise video in design, there are a number of established traditions upon which to draw. Although they all aim at designing more useful products, they each have their own theory and methods, and their attitudes towards users differ as well. Design ethnography uses video to study the daily practices of (potential) users and to communicate findings to designers. It sees users in the role of informants. Participatory design involves users as participants in a design process, and video is employed to document design discussions and activities. Usability studies identify user problems by simulating use situations in a lab experiment. Usability specialists use video as an instrument to document the reactions of test subjects for later analysis and for communication of the results. Scenario-based design uses video as a medium for creating and telling stories of future user interactions with imaginary products. Users are often seen as actors in the product stories. In this section we will briefly discuss the practice of each tradition.

Design ethnography

Design ethnography (or design anthropology) is an emerging field that integrates the study of people into a design process. Design Ethnography builds on the long history of social and cultural anthropology, which employs field study as the traditional method for the careful study of activities and relations between people in a complex social setting. Ethnography refers to the *description* of people, and it aims to describe the cultures, activities and traditions of indigenous people from the point of view of the community members.

With the advent of film and later video, ethnography was supplemented or even replaced by ethnographic films that represent foreign communities in a vivid and visual manner. Films typically cater to a broader audience than written ethnographic accounts. Pink (2001) points out that ethnographic videos communicate a broad set of issues in parallel. For instance, the video recordings that she recorded during a study in Guinea Bissau in 1997 illustrated how the activity of weaving unfolded. The video records also communicated how the friendship between the researchers and informants developed during the study. Furthermore the interconnected comments of the people on the videos present an account of cultural beliefs and attitudes (Pink 2001, p. 149). Ethnographic video presentations vary from raw video clips to professionally edited documentaries, depending on the case. Video materials are also utilised as still pictures and transcripts in descriptive passages for book journals.

Design ethnography differs from traditional ethnography in that it studies the culture of the potential *users* of the technology in focus. For design, a written ethnographic analysis plays a less pronounced role than in ethnography in general. The goal of design ethnography is to provide sufficient understanding of the studied practice in order to discover new opportunities for design. Thus, rather than handing over finished explanations, the ethnographer in a design team (or the designer employing ethnographic techniques) tries to convey the concrete richness and ambiguity from the field in a form much more open for interpretation, as this will spur the discussion on change.

When compared to traditional ethnography, design ethnographers maintain the point of view of the people studied, but they take a more narrow view on how people interact with technology. Design ethnography is characterised by a fair, rather than exhaustive, understanding of the participants' practice. It produces a "thin" description of the culture where traditional ethnography produces a "thick" one. Design ethnography is done at great speed compared to traditional ethnography, because month-long studies would seldom make sense in commercial product development.

The core significance of design ethnography for design is the help that it provides in focusing the design on the "right product". This is only achieved by being open to, and paying enough attention to, the richness of the real social settings. Robert J. Anderson, social scientist and former Director of the Rank Xerox Research Centre at Cambridge, emphasises the value of ethnography in enabling designers to "question the taken-for-granted assumptions" in the conventional ways to do design (Anderson 1994, p. 170). Design ethnography seeks to sensitise designers to the relevant real-life ambiguity. The extended use of video might render the need for formal descriptions obsolete as designers can rely on video to inspire their imagination. A description in itself – explaining the past – may not have value once the product has been conceived and brought to market.

With its capability for prolonged observation, video can reveal behaviour that would otherwise remain undiscovered. Leaving the video camera to record while the researchers leave the scene enables rather unobtrusive studies. This method is useful in situations where the participants' work is sensitive to disruption. Video provides access to some scenes that would otherwise be beyond detailed analysis. Suchman and Trigg (1991) used a stationary camera for exploring the use of tools in particular spaces in an airport. These videos enabled a close and unobtrusive study of the interactions related to the use of flight-related paper documents (Suchman and Trigg, 1991).

The methods of design ethnography continuously develop as new technical opportunities arise. Mobile terminals with image, video and voice recording capabilities facilitate new methods. For example, *digital experience sampling* is a new way of gathering information. Participants document their behaviour at certain intervals with a digital tool, and the data is transmitted to a server for the designers or researchers to analyse. Platforms such as *Mobile Probes* (Hulkko et al, 2004) have been developed to enhance the possibilities to conduct interactive ethnographic studies with mobile devices. Digital tools such as digital cameras, PDAS, laptops, virtual collaboration sites or other technologies are increasingly being used to record, transmit, edit and present the information about the users' reality.

Interaction analysis

The possibility to review video recordings from the field drastically changed the way of analysing ethnographic material. As handwritten field notes are personal, analysis too has traditionally been individual. Video captures what happens in the field with sufficient richness to allow different observers to contribute with their interpretations. This cuts down the analysis time while enabling some breadth for the interpretation.

Video also adds details to the field records. Handwritten field notes cannot be anything but an *ad hoc* account relying on the writer's reconstruction of events. Human activities unfold so fast that it is impossible to capture their complexity by observation alone (Jordan and Henderson, 1995). With the option of replaying a sequence over and over again, slowing it down and pausing it, video is a remarkably useful tool for analysis. It has enhanced the range and precision of the analysis of real-context interaction remarkably, and the detailed and close-to-reality nature of the video data "provides some guarantee that the analytic conclusions will not arise as artifacts of intuitive idiosyncrasy, selective attention or recollection" (Whalen *et al.*, 2004, p. 3).

Conversation analysis was originally developed by Harvey Sacks in the 1960s at the University of California. He worked from the assumption that spoken language is *designed* by people for the particular situations in which it is used and that there is "order at all points" in *talk-in-interaction* (Hutchby and Wooffitt, 1998). Sacks' work on conversation analysis rested in large part on his original and iconoclastic way of thinking, but also on the newly available recording technology, which could produce detailed records of human conduct for close analysis. The multi-disciplinary conversation analysis is grounded in linguistics, sociology, anthropology and psychology, and it aims to reveal the tacit, organised reasoning procedures that inform naturally occurring talk. The analysis is based on precise transcriptions from audio and video recordings, and it commences without prior theoretical assumptions about what is to be found in the data. The aim is to discover how people understand each other and take turns when talking. Like all ethnographically-oriented work the analysis focuses strongly on the context where the conversation takes place.

Building on conversation analysis as well as on more recent theories on situated action, especially that of Lucy Suchman (1987), Brigitte Jordan and Austin Henderson developed *interaction analysis*. This is based on Interaction Analysis Lab sessions in which researchers from various disciplines are brought together to view and discuss selected video recordings (Jordan and Henderson, 1995). The method is described further in Chapter 3.

Activities in real life unfold at an incredible speed. There are too many processes going on, the active structures are too complex, and there are too many aspects to focus on, to create a detailed understanding of what happens when people act. Hence, some reflection is necessary for building a shared understanding of what actually occurs and for making the understanding actionable for a design team. Interaction analysis also adds a social dimension to the analysis, which is necessary for design, as most products relate to several people, at the same time or at different times.

The growth of both design ethnography and interaction analysis is tightly coupled to the development of video recording: video allows a design team to learn more in short, condensed field studies; it provides a resource for collective analysis; and it may replace the written ethnography to better inform the design process.

Participatory design

Participatory design developed in Scandinavia in the 1970s and 1980s to empower workers to influence the technology with which they were required to work. It was originally a political movement, where systems developers strove to increase democracy in the workplace in collaboration with labour unions (Ehn and Kyng, 1987). In the late 1980s participatory design caught interest in the USA and has since grown and changed from a political into a pragmatic approach to design. Inviting users to participate in design has positive consequences beyond giving them the power to affect the tools that they will be using. Participatory design is observed to have increased the ownership of the design ideas through increased knowledge about the development as well as made adoption of the new solution more fluent (Blomberg *et al.*, 1993).

When it comes to the practical organisation of participatory design activities, the main challenge is to support fluent collaboration between designers and users. As the participants typically come from different professional traditions, there are barriers to be crossed. Participatory design is typically organised around workshops where participants spend time building things together and discussing. Participatory design researchers have developed a range of methods that make design activity accessible also to users. Simple materials, photographs, stories, acting, game-playing and mock-ups are utilised to give users with no design training a chance to contribute by giving them ways to express their needs and enabling them to give feedback and to suggest improvements. The methods are often playful, fast and inspiring in order to make design engaging for the participants and ensuring their future participation.

One such method is PICTIVE (Muller, 1991). This is a collaborative game for facilitating user participation in the user interface design of computer software. The method has refined the use of simple materials for paper prototyping of screen interfaces. In a PICTIVE session the participants work with ready-made materials, such as paper buttons, menus, and pop-up windows, to design a computer interface. A video camera is used to record how a user interacts with the design both to document a concept and to detect flaws.

In addition to enabling communication a central challenge in participatory design is to keep the collaboration grounded in the use context at all times. Just as prototypes essentially solve the problem of communicating technology knowledge to users, methods are necessary to make the users' domain knowledge available in the design discussions. Participatory design typically relies on stories from real life and scenarios of the future. There are many examples of acting out scenarios in user workshops (see, *e.g.* Bødker and Iversen, 2002), where video is employed to document the scenario for future study or reference. In recent years there has been a trend of designers moving out into the users' environments and acting out scenarios with the users to obtain immediate feedback on new design ideas. For instance, Thomas Binder (1999), when employed with Danfoss, asked process operators to act out familiar maintenance routines, albeit improvised with simple mock-ups of digital tools. In such sessions a video camera acts not simply as a documentary device, but helps initiate collaborative authoring between users and designers.

Scenario-based design

Scenarios – when applied in product design – are stories about potential future use situations with the design solutions. Scenarios were originally used in military and business contexts for imagining alternative states of the future in order to better prepare for the one to come. Scenarios are widely

used in the design of interactive information systems, consumer appliances,
services, *etc*. Scenarios are found to be useful throughout the product development lifecycle for creating and presenting ideas, discerning user requirements, and evaluating ideas and prototypes.

One of the key strengths of scenarios for design is their ability to embed
the use context into the presentation of a product. With the use context provided, the audience of a scenario is able to evaluate to whom the product is
suited, where it is useful, the objectives the product supports, and how well
it functions for its purpose. A scenario may present interactions at any level
varying from interactions with particular functions of a product to lengthy
human interactions in a socio-cultural context. The scalability in format and
detail has allowed scenarios to be utilised in a wide variety of projects.

Video as a means of creating scenarios has been inherited from cinema
movie-making. A scriptwriter builds an imagined reality, which the director, with the help of the movie crew (actors, cameramen, *etc*.), realises as a
movie. In a similar manner large companies produce movies that illustrate
the future as they imagine it. For example, SunSoft's visionary film "Starfire"
(Tognazzini, 1994) described a future with curved displays and advanced
means of interaction that were imagined to be possible in 2004. The reality did turn out to be rather different from what SunSoft designers foresaw;
however, the scenario movie supported the broader aims around it, especially
in promoting the company brand. Professional video scenario productions
have also been presented by companies such as Apple, Hewlett Packard, AT&T
and Phillips. These video scenarios in a way replace the need for functional
prototypes that provide people with the overall experience of the system in
fluid action. Such scenarios are good for raising debate on what may be a
desirable future, thus paving the way for making decisions on partnering
and possible projects to launch.

However, effective video scenarios do not need a movie budget. During the early design phase a sketchy use of video is well-suited to exploring
new ideas. Binder's (1999) improvised scenarios, for instance, were simply
shot with a handheld camera and a coarse foam prop, albeit in a real plant.
Mackay, Ratzer and Janecek (2000) utilised a technique called "video brainstorming" to enable designers to present ideas in a more vivid and memorable way compared to writing the ideas on paper. The authors acted out the
ideas using simple mock-ups in front of a video camera.

The large variety of ways in which video is utilised in exploring, evaluating and displaying ideas makes it problematic to see clear-cut categories,

and naming conventions vary considerably. For example, the heading "video prototyping" is associated with such diverse uses of video as video-mediated presentations of a mock-up user interface in a participatory design session (Mackay *et al.*, 2000), large scale scenario movies (Tognazzini, 1994), designers' acted-out presentations of ideas with mock-ups (Mackay *et al.*, 2000), and stop-watch animated presentations of user interfaces (Vertelney, 1989).

Usability studies

Studying usability is a rather new practice dating back to the early 1980s, when software products that were formerly used by computer professionals became available to the mass market. Many software products were found to be too complicated, and usability studies were developed to make the product easier to use, more efficient, less error prone and more satisfying for people. Usability studies aim at improving usability by detecting usability problems. The practice developed from scientific laboratory testing, which originally stemmed from experimental psychology, and moved to more practical methods that better serve the needs of industry. Current usability studies comprise a variety of methods, such as heuristic evaluations, usability walkthroughs, and usability tests with representative users (Nielsen and Mack, 1994).

In usability tests the users are instructed to accomplish defined tasks with a mock-up, a prototype or a finished product, usually in a laboratory setting. The users are asked to think aloud during their interaction with the product to allow the researchers to capture their thoughts in addition to their actions. The test is documented in detail with tools such as data forms and video. A usability laboratory is usually equipped with several video cameras that are pointed towards different areas of interest, such as the user's face and the user interface of the product. With dedicated software the data from multiple sources can be combined for later evaluation, and annotation of the data is possible while the user is doing the tasks.

Usability specialists use video to gain a better view of the user interactions without disturbing the process and to generate a detailed record for later studies and presentations. In particular, usability professionals use "highlight tapes", which are edited video movies showing the central usability problems, for effectively communicating the findings of the usability tests to project participants and managers.

However, video is a challenging tool for usability use because video material is perhaps the most demanding type of data for usability analysis. According to Jacob Nielsen (1993), one of the pioneers of usability studies, the analysis of videotapes takes three to ten times the duration of the original activities. When the findings are edited into polished highlight tapes, the time requirement is expanded to even 60 times the original duration of the activity (Dumas, 2003). These figures render the traditional use of video for usability studies rather questionable. However, new software tools enable usability experts to annotate the usability test data in real time during a usability test. This allows the creation of a video of the key usability problems in an instant and is likely to increase the utility of video in usability tests in future.

Making video efficient for design

The value of video for design projects depends heavily on the approach taken. Two metaphors are proposed here to highlight the main roles of video for design. The first – *video as designer clay* – explores the productive side of design: what the video movies represent, and what designers are able to express using video. The second metaphor – *video as social glue* – helps understand how the video equipment and the situations of shooting, editing and showing of video support the social process of design: how people collaborate and develop ideas together.

Video as designer clay

Industrial designers use malleable materials – like designer clay – to model the shape and appearance of a new product. With the clay a designer can quickly build and modify alternative versions, and it is rather easy to communicate product design ideas with such a concrete material. Designer clay has a special ability to stay mouldable – unlike ordinary clay, which hardens. This quality allows the designer to come back to the concept even after some time and modify the shape based on the new understanding. What if designers had a similar type of clay that would allow them to sculpt the less concrete aspects of product design? With such a tool they would be able model much more abstract concerns.

- ▶ Who are the expected users, what do they do, and what do they like?
- ▶ Which core themes should the conceptual design pursue?
- ▶ How will new product proposals fit into the user's environment and practice?
- ▶ How will users interact with the new product?

Video is such a medium. It can capture activities as they unfold in time; it can portray the personality and feelings of people; and it can show a fictitious future. To emphasize this quality of video we prefer to talk about video as clay rather than data. Data carries the notion of objective research, of truth that cannot be questioned. Design challenges are open without a limited set of right (or true) solutions, and approaching design from the point of view of truthfulness presents a misconception of the pursuit. Clay, rather than data, can be shaped by a designer until he or she is satisfied with the form. Moreover there is a certain intensity to the shaping itself: the very process of moulding is a process of coming to an understanding of the conditions and possibilities of a particular design.

Video captures the temporary aspects of the world around us; it lets us preserve and study how life unfolds within our focus of attention. Video material – as clay – allows the designer to then mould interactions as they unfold in time and space: both the interactions between people and between people and technology. The designer can sculpt the interactions as they appear today, or as one may envision them in the future.

With ordinary designer clay, the industrial designer communicates in very concrete terms to anyone he or she chooses to share ideas with. The form of the product is obvious, even though much of the inner functionality is not apparent. In the same way video is a powerful and very concrete form of communication. The interaction designer can involve others in discussing interaction, even though it is not available as a physical object. Video materials allow the interactions to turn into catalysts of a dialogical learning process rather than as static sources of objective user data. This book demonstrates how designers can propel the design process forward through formation and transformation of a particular kind of presentation – video artefacts.

Video artefacts may turn into mere by-products of the knowledge acquisition process of design, or they may gain a high value in driving strategic innovation efforts of businesses. An open attitude towards the use of video is crucial to its utility. Video design artefacts – like pieces of art – may become valuable in themselves.

Video as social glue

The design of complex interactive products is a social process as much as it is the craft of producing something new. A design team is typically composed for the project, and team members who may not have worked together

before need to unite their various professional competencies to make the
project a success. Henry Dreyfuss (1967, p. 22) stated:

> He [the industrial designer] must be part engineer, part businessman,
> part salesman, part public-relations man, artist, and almost, it seems at
> times, Indian chief.

The designer does not need to be a transdisciplinary wonder man, but has
to be able to talk and interact with people with various skills and different
backgrounds. Moreover, it would be helpful if collaborative efforts were arranged in ways that enable mutual participation and engagement. Used with
respect for human relationships, video can bring people together around
design activities and relevant discussions – it can work as the "social glue"
between the stakeholders in a design process. Video may help bind together
a multidisciplinary team and close gaps between the design team and users,
and between the team and the rest of the company. This is important, since
collaboration is typically strained by the different backgrounds, professional
languages, and interests of the participants in the process.

What about the moving image that creates this effect? It is the *concrete
richness* of video recordings, for one. A video presentation of real-life activities is capable of displaying a world that the audience is familiar with and
can make observations about. As in real life there is a myriad of perceptions
possible. The viewer can relate to the video on many levels and focus on
many different issues: Who is there? What is the environment like? What
activities are there? What tools are used? How are the people on the video
feeling? What are they saying? Despite the detail of video presentations,
the viewer has to keep in mind that video records do not convey an objective image of reality. Someone has decided where to point the camera and
when to record, and the presence of the camera operator and camera often
affects the setting. Nevertheless, video is the medium that conveys most of
the detailed richness of a real setting, as compared with text, photos and
audio recordings.

Despite its concrete detail video is *ambiguous*. It allows varied interpretations – just as in real-life situations. Viewers can decide on their own what to
believe and why. Some interpretations may be built into the material through
the process of authoring and editing, but as long as we do not move into the
realm of million-dollar cinema productions – where the director may control every detail in an attempt to convey a particular experience – designers'

videos are likely to leave extensive room for discussion and multiple interpretations. For most people it comes as a pleasant surprise when they realise
how many other observations and interpretations in addition to their own are
possible – at least once they get over the painful revelation that their way of
seeing things is not the only one, and may not be the "right" one. This clash
of views immediately triggers discussion: "Why does the energetic product
manager see things differently from the empathic engineer, or the cautious
physiotherapist?" In particular, video about users and future products works
to trigger discussions across disciplines and interests.

The acts of making and watching the video presentations are often *enjoyable* and *exciting*, not least for the participants acting in and capturing the
videos. In the case of video scenarios, making a "movie" is a new experience
for many. Once a group of people have taken their positions and prepared
the materials for the next shot, the atmosphere is often filled with exhilara-
tion – something quite different from the writing of reports and creation
of slide shows. In some cases, the making of the video artefacts extends to
creating manuscripts and to collaboratively editing the videos. Watching the
moving picture is often a captivating and more memorable experience than
the reading of reports and summaries.

With this book and the related video presentations designers can learn
how to use the video camera to provoke design actions: both by triggering
users to show what they do and want to do, and by triggering the design
team to act out, and to concretise ideas of new products and interactions.
The intentional editing of video collages, portraits and scenarios can invite
users into discussions about present and future. Designers can also employ
carefully crafted short movies and highlight tapes to instigate debate and
change in a corporate organisation.

The Interactive Kitchen design case

Kitchens are familiar to all of us; developing radically new ideas for a kitchen
might thus turn out to be very difficult and require a shift in perspective.
Video helps create an appropriate distance from the kitchen while offering
a grip on the details in order to register something beyond personal experience. Video scenarios fuel exploration and facilitate communication of new
ideas. This case study, which was done in collaboration with researchers and
students in Denmark, Germany, Finland and the Netherlands, shows how
the various design video artefacts are utilised in practice. It also illustrates
some of the social aspects of the design events with video.

The Interactive Kitchen design case had a very open starting point: "Is it possible – and desirable – to introduce interactive technology into home kitchens to support what people actually do?" "What design concepts can we imagine, and how do they then change the practice of cooking in the kitchen?" The video examples on the DVD stem from a range of activities relating to kitchen design, some in Finland, and some in Denmark.

In all, there were seven two-hour studies. The designers completed rather open, ethnographic-type observations, using a video camera to learn how people cook a meal in their own kitchen. The studies did not focus on any particular product but aimed at capturing a picture of the whole, of how people cook in their homes. The video material from the kitchens, *i.e.* the field study *video footage*, forms the basis for the subsequent design activities. The DVD contains an unedited excerpt from the field studies of Tanja, who is cooking lentil soup. To provide a feel for the atmosphere in a field study we have included this unedited portion from one of the raw video recordings. It is rather long, so students may use this to train observation and editing.

Video footage
Tanja in an atelier kitchen
7'26"

Already during the field study, there was a complex authoring process going on in "modelling" the recording: the designer had to choose where to stand, where to point the camera, and what to record based on some understanding of what was perceived to prove interesting for the project. Similarly, the people in front of the camera tried to adapt to the brief given by the designer about recording. People tended to

adjust their actions – in subtle ways, perhaps – to better show to the camera what could be interesting. Traditional HCI researchers would probably scream "obtrusive camera!" at this point and claim that the very presence of the camera intrudes on the activities and thus blocks the "objective truth". The camera, as well as the presence of the person observing, certainly has an effect, but we claim that this is not a disadvantage: rather the contrary. By providing a solid introduction to the project and to the purpose of the video study at the beginning of the kitchen visits, the designers were able to engage with the reflective and creative capability of the people studied. When people knew what was happening and understood what was expected of them, they were able to mould the best possible image of the practice together with the designers, and turn video into "designer clay" already during the filming phase.

Figure 1.5
A still image from the video footage from one of the kitchens

The camera also had a social effect. The activity of filming was a sign that someone was interested in what the person was doing. Often this was perceived as pleasant by the participants. Achieving this required careful preparation and explanation of the purpose and use of the video recordings, as well as giving the participants control over the event. When the purpose is clear, video can "glue together" the designer and the user in an effort to create something collaboratively – an accountable video recording of what goes on.

In the editing stage, the designers used the raw footage to create "video artefacts". Editing allows a myriad of approaches, and at each point of the edited material there are innumerable possibilities to pursue. Based on what the designers perceived as interesting, choices were made on scenes, sequence, editing points, timing, and rhythm. There are no objectively optimal solutions at the moment of editing, because the value of the video artefacts cannot be known until they are created and displayed. The artefacts functioned on the level of interpretation and communication and – in a way – the designers "moulded" their understanding into the new video artefacts, which then provided new ground for further learning.

Video story
**Tanja
prepares
lentil soup**
6'43"

In the kitchen study, the designers utilised three basic forms of video artefacts. First, *video stories* were extracted and studied to help a design team to explore how things happened at the user site. This provided the team with knowledge on the flow of action and conflicts in the current practice, and thus enabled them to identify new opportunities for improvements. In

Figure 1.6
The video story shows how the activities in the making of lentil soup unfold

kitchens people tended to juggle several tasks simultaneously. The exercise of pinpointing each particular flow of actions was quite a rewarding one. During the study it became clear just how differently people went about preparing dishes. Some went by the recipe book; some went by experience. Some washed and cleaned as they went along, some waited till the end. Some seemed to have a calm flow of activities, others were more dynamically paced.

Secondly, the studies were made in a manner that allowed designers to create *video portraits* of the home cooks. The videotapes had two types of content: interviews and observation. The observation videos showed what the participants were doing and what they looked like, and the interview presented what they said and thus communicated what they thought about things. Editing the portraits required a bit more observation and editing skills than the video sequences, as it required the designers to grasp the essence of

1 *Video
in design*

Video portrait
**Tanja the
gourmet**
3'00"

the practice and personality of each person, and then to communicate this through scene selection and editing. Indeed the persons studied exhibited very different personalities in how they cooked, in their values and preferences. Some aimed at being efficient; some strove for a balance in taking care of children and family, and some would never use prepared foods.

Figure 1.7
The opening
scene of a
video portrait

Thirdly, cooking at home is familiar to everyone; therefore, a fresh way to look at it was important in order to push ideation into radically novel tracks. In addition to the analysis of the activity flows and descriptions of people's values, designers edited rough *video collages*, juxtapositions of rather raw video clips. Through interpreting and grouping video clips together in a video card game (the method is explained in Chapter 3), designers found new metaphors (or themes), such as "the skilled knife", which pointed designers to ponder why everyone seemed to use the biggest knife for all the tasks, from cutting carrots to peeling garlic to splitting olives.

The discussions on the themes, such as "the social recipe book" and "measuring with the mouth", discovered through collaborative interpretation of the video clips, helped the design team to lift off from the conventional ways of seeing people's performance in kitchens. When seen against the background laid by the site visits, the making of the video portraits and stories of various courses of activities also helped to raise relevant new questions, such as, "Where did they need the recipes, and how were they used in 'social' cooking?" Questions like these led to the discovery of new opportunities – or challenges.

The initial development of new ideas began with an improvised puppet scenario workshop. The stage was set up with a cardboard model of a

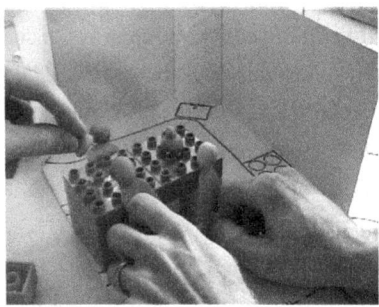

Video collage
**The skilled
knife**
3'34"

Figure 1.8
Video collage
on various
ways of using
the big knife

kitchen based on the user studies, and a video story was displayed to the multidisciplinary team to ground ideation into the real practices in kitchens. The team then started to craft ideas by brainstorming with the blocks of cardboard, scissors, pens, and some children's clay. With these materials the designers built the environment in which to act the ideas out collaboratively. The ideas thereby became expressed through illustrating how they would function. The play with the puppet figures was a lot of fun, and it helped to escape the fixated ideas about how kitchens used to be.

The exploration of new means to facilitate cooking as a social practice – a user need identified in the study – developed into one of the major tracks in the project. The ideas covered a number of ways to transform, *e.g.* microwave ovens into new kinds of tools to enhance the sociality around kitchens. The other major track focused on constructing new tools that incorporate modern information and communication technologies.

Video
scenario
**Puppets in
the kitchen**
0'59"

The ideas were developed into concrete mock-ups of actual size, and they were placed into settings with real people – as themselves – experimenting and improvising with the mock-ups. These improvised scenarios were captured with a video camera and helped to ground the discussions on how to improve the final designs.

The track that focused on building product concepts utilising information and communication technologies created ideas for new kinds of

Figure 1.9
Designers
improvis-
ing ideas in
a cardboard
environment
with puppets.

appliances that may be situated in a kitchen setting. Ideas that explored opportunities to check e-mails, improve the planning and coordination of cooking, and to communicate real-time with other cooks were among the wealth of issues discussed. One of the ideas was the dedicated email reader, a tiny box featuring a turning dial on its top. Through twisting the top the family members could choose if they would read e-mails, browse slide presentations, listen to voice-mails, *etc*. The wireless reader could be placed anywhere in the home, for example, on the kitchen table, where it would become a social gateway to all the networked materials.

A full-scale mock-up was also crafted to enable testing the dedicated e-mail reader in homes. The videotaped field test was helpful for capturing not only the reactions to the idea but also visual material, creating graphical

1 Video in design

Video scenario
The social microwave
1'23"

Figure 1.10
Improvising with a lifesize mock-up of the new microwave

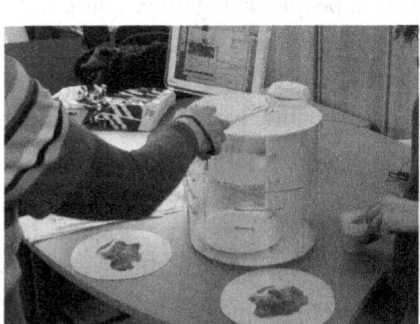

presentations from the video stills as illustrations for posters. The field test also provided insight into where the product would practically be located in households, as well as how different people would be able to control that form of a product.

About methods and the structure of the book

Traditionally, in engineering, methods have been thought of as sequences of actions. In this way, one would get to a result by simply following instructions. This is derived from a production philosophy in which goods are produced in a sequence of operations with clearly defined substages. However, if we understand designing as a predominantly social activity there is much more going on in design practice – in what people do together. There is negotiation, collaboration, debate, conflict, and other social action.

This book promotes the idea of methods as a set of activities embedded in a particular environment, with participants, materials, tools, and a general direction, or a goal: what to achieve. Method may be understood as organising a party. People do not plan precisely what the guests need to do step-by-step, but they put all their energy into organising the space, the lighting, the food, the decoration, the songs and the music. If the party organiser achieves the right atmosphere, the participants will make sure that the party is a success.

This book describes 16 methods that span video activities in the entire user-centred design process. Rather than following a uniform template, they describe methods that vary significantly in structure and use. In some, the sequence of activities is important; in others, it is the skill of collaborating with people, making sure that staging and framing is in place. Some of the examples manifest the method by describing the results.

The case stories aim to provide a contrast to any rigid method description. Where the method descriptions aim to outline a simple and understandable practice, the real-life examples illustrate how life mixes, conflicts with and combines methods into new formats that provide successful results in the particular conditions where the videos are created and used.

The accompanying DVD disc contains excerpts of these actual cases. These give concrete images of the kinds of results, and intermediate artefacts, that designers have produced and utilised to inform and inspire design in both academic and industrial contexts.

Figure 1.11
The dedicated email reader. The field test of the full-scale mock-up and the final poster. Design by Kim Aagaard Holm

One operator analyses the wat

2

Studying
what people do

JORIS IVENS

"The film's art begins when you choose where to place the camera."

2

Studying
what people do

When designers create new products for people whom they do not know, they need to engage in activities that render the use context visible to design. This can be done with video studies of users, where the video camera is employed as a tool to construct relevant material that both informs and inspires design. This, however, is not the full story as to why designers benefit from using a video camera. Insightful use of video in user studies turns the enquiry into a constructive dialogue about what is seen and how people see it. Video studies foster the collaborative construction of a design-focused understanding of the users' reality.

Traditionally ethnographic research has sought to describe the cultures studied in a detailed manner. From such descriptions ethnographers have identified patterns and built theories that have the power to explain the phenomena on a more general level. In contrast, designers with a video camera look for facts and inspiration, and they strongly affect other people's reality, impacting people on both sides of the camera. Essentially, video provides a means to engage different people in a collaborative learning process. At times the use of the video camera may present a credible "excuse" for mingling around the user site and observing the activities. Nevertheless, rather than seeing video-based fieldwork as a means to collect rich user data, this chapter outlines a practice of co-authoring video materials with users and of framing design challenges in novel ways.

The ethnographic camera

An increasingly popular approach to studying users for professional design is design ethnography. Ethnographies are written descriptions based on fieldwork, where an ethnographer participates in people's daily lives for an extended period of time, observing, interviewing and collecting data within the focus of the study. The primary method associated with fieldwork is participant observation, *i.e.* being there in the natural setting and observing what goes on. The time spent in the field varies from a few months to several years. Ethnographers tend to build close personal relationships with their informants, to the extent where ethnographers talk about "going native".

Video use in design ethnography originated in the work of visual anthropologists, who began to utilise video in the 1980s. They praised the convenience, economy, durability and utility of video compared to film. Video made it possible to record people's activities continuously for several hours and enabled reviewing the material instantly after capturing. The capability of instant review enabled ethnographers to gain more detailed views on the activities captured on video with the informants (Pink 2001).

What is "practice"?

During the last two decades a transition towards understanding "practices" has taken place both in the discussion of academic knowledge as well as in the theories about and methods for user-centred design. What is this "practice" that designers need to study in order to design products that fit? Practice is something people construct themselves, which becomes part of their identity. Etienne Wenger (1998, p. 6) has shown how people fundamentally learn in organisations:

> *Workers organise their lives with their immediate colleagues and
> customers to get their jobs done. In doing so, they develop or preserve
> a sense of themselves they can live with, have some fun, and fulfil the
> requirements of their employers and clients. No matter what their
> official job description might be, they create a practice to do what
> needs to be done.*

A practice is inherently bound to the local conditions of context. Indeed, it is not possible to understand practice without understanding the local conditions, argues Andy Crabtree (1998), a social scientist who has studied the value of ethnography for systems design:

> ...enacted practice is highly localised, contingent, and (above all) subject to continuous enquiry and discovery for practitioners themselves in the course of work's accomplishment. Thus, enacted practice is, to some significant extent, intransigent to explication in alternate contexts; hence the need to "take a closer look".

Moreover practice is fundamentally social by nature. Lucy Suchman (1987) identified four main reasons why previous theories and methods were not sufficient to grasp reality for design in a suitably sensitive manner. First, mutual intelligibility of interactions is always the product of *in situ*, collaborative work. Second, the general communicative practices that support that work are designed to maximize sensitivity to particular participants, on particular occasions of interaction. Third, face-to-face communication includes resources for detecting and remedying troubles in understanding as part of its fundamental organisation. Fourth, every occasion of human communication is embedded in, and makes use of, an unarticulated background of experiences and circumstances.

For example, Hughes *et al.* (1994) observed how ubiquitous technologies for networked and distributed activities generated unforeseen effects in collaborative practices, because the widely employed methods for eliciting systems requirements were unable to address the social organisation of work. Moreover, practice is not stable. Hughes *et al.* (1994, p. 435) describe how "human beings have an extraordinary ability to 'make do' with the technology with which they are provided". Human practices evolve rather rapidly in response to changing conditions, for instance as a result of new interactive products becoming available – regardless of it being work or leisure. Hence, for the study of these phenomena designers need methods and tools that enable them to address the processes of the social organisation of action in people's native settings.

How video helps

Paul Dourish, a researcher of computer-supported cooperative work, contends that people may not actually do what they say they do, or they may do many things that they omit when asked to talk about what they do. Often it is the case that "the ways the work gets done are not the ways that are listed in procedural manuals – or even accounts that the people themselves would tell you if you asked" (Dourish, 2001, p. 19). There are numerous reasons for this: first, when things begin to happen automatically, conscious awareness is not necessary. Actions become automatic, slip to the background of consciousness, and may thus escape any attempt at listing or recognising them without having the context to support recall. Second, formalised practices are basically always too rigid to represent real social behaviour in people's everyday settings. The former head of Rank Xerox Research Centre, Bob Anderson, explored how the issue propagates to product specifications. He argued that "requirement specification", which refers to the formal description of product properties, cannot address the details of the dynamic and complex everyday reality in which the designed products ultimately need to fit, and another approach (namely ethnography) is therefore necessary.

> *What the user is held to know about and to orient to in the daily routine of their workaday world is the practical management of organizational contingencies, the taken-for-granted, shared culture of the working environment, the hurly-burly of social relations in the work place, and the locally specific skills (e.g., the "know-how" and "know-what") required to perform any role or task. Formal methods of requirements capture, or so it is supposed, are incapable of rendering these dimensions visible, let alone capturing them in the detail required to ensure that systems can take advantage of them. In our view, ethnography is at least a method that will provide access to these dimensions.* (Anderson, 1994, p. 154)

Ethnography is becoming commonly acknowledged as an apt approach to building the design understanding of people's real social practices at an appropriate level. For example, Hughes *et al.* (1994, p. 432), who reviewed experiences from numerous ethnographic studies, affirm that:

> *What the ethnography especially provided was a thorough insight into the subtleties involved in controlling work and in the routine interactions among the members of the controlling team around the site: subtleties which were rooted in the sociality of the work and its organisation. The vital moment-by-moment mutual checking of "what was going on"*

by the various members of the team had been missed by earlier cognitive
and task analytic approaches to describing controlling work.

In particular, *video ethnography* has proven an invaluable means to address
the details of everyday activities. For example, the outstanding work by Chris-
tian Heath and Paul Luff (2000) to study technology in social interaction
was completely grounded in the detailed analysis of video recordings. Video
is *the* tool to capture the production and coordination of real-life activities in
their native settings. According to Heath and Luff video has three qualities
that make it especially suited for the analysis of interactional organisation
of workplace activities: first, video provides access to the details of talk and
visual conduct, enabling a detailed scrutiny of the activities, if necessary, with
slow motion; second, video recordings enable researchers to share the data
with colleagues and thus enable discussion on the materials on which the
analysis is based; and third, video enables the public display of the findings,
thus subjecting the findings to public scrutiny. Based on experiences in a
design project preceded by an extensive ethnographic video study, Crabtree
et al. (2002, p. 269) also promote this capacity of video:

> *In practical day-to-day details of "getting activities done", video ethnog-*
> *raphy furnishes investigators with fine-grained and phenomenally intact*
> *in vivo recordings of everyday family life. In contrast to a mass of notes,*
> *anecdotes, vignettes, and disembodied conversations which character-*
> *ize traditional ethnography, video footage becomes the primary resource*
> *enabling direct investigation of the domain.*

For designers, video is capable of capturing activities in a manner that
holds the contextual aspects intact rather than delivering de-contextualised
generalisations of the issues encountered. However, despite these benefits,
video ethnography is highly problematic. The main problems relate to the
relevance, scale and quality of the studies.

Once descriptions of social interactions are made, they turn into frozen
artefacts merely depicting history. Any change introduced to the scene is
likely to affect how things become accomplished. Social interaction with
technology is dynamic and responsive to the technical interventions that
designers create. Hence, it is questionable how much designers need to
know about current practices in order to facilitate a new technology-mediated configuration of future activities.

Ethnographic studies for design have largely focused on rather constrained areas, such as control rooms. Such a study enables the detailed scrutiny of the micro interactions within that space. However, when the scale is expanded to functions across departments and organisations, the difficulties in capturing the details of interaction will explode, and, moreover, the likely relevance of the diverse details of micro interactions on the whole will abate. According to Hughes *et al.* (1994, p. 431):

> *Scaling such inquiries up to the organisational level or to processes distributed in time and space is a much more daunting prospect in raising issues of depth and representativeness.*

In addition to these issues, what designers will face are scarce resources for conducting ethnography. The main resource design ethnographers do not have is time. In industrial organisations user field studies need to align with the rapid product development cycles of a few months. Hughes *et al.* (1994, p. 431) continue that:

> *As one of our computer science colleagues expressed it, ethnography is a "prolonged activity" and in the context of social research can last a number of years, certainly time scales which would be considered a joke in software engineering. Added to this are the problems, noted earlier, of communicating ethnographic findings to designers. The output of ethnographic analyses are typically discursive and lengthy, looking nothing like the blueprint diagrams which are de rigueur in systems engineering.*

Design ethnographers count their field studies in days rather than months. In response to this, a research group at Lancaster University's CSCW Centre introduced the term "quick and dirty ethnography" to describe the type of studies required in development projects (Hughes et al, 1994). Such studies are characterised by a fair rather than exhaustive understanding of the studied practice. Some of the techniques for conducting "quick and dirty" ethnography are presented later in this chapter.

Videotaping reality?

Designers need to understand the users' reality. What, though, is reality? How can someone say something about what reality is? Anderson (1994, p. 155) warns us that:

*...the supposition that ethnography conveys an overall impression of
"what life is like" or "tells it as it is" is profoundly mistaken.*

The question of reality is one that philosophers have debated for millennia.
When we take a constructivist position, *i.e.* acknowledge that the influence
of people's subjective and shared perceptions of reality constitute their consciousness of it, we must accept that no one can state purely objective truths
about reality. Even the most purist documentarist who captures real life with
the film camera acknowledges that movies are far from objective. To understand this we need to go a bit deeper into the discussion.

During the last century documentary movie authors developed theories
of presenting claims about *reality*. When technical advances in cinema technology revolutionized documentary making in the 1960s, portable cameras
and audio recorders enabled documentarists to descend into and move
with people's everyday activities. This approach was coined direct cinema
(in Canada and the U.S.), *cinéma vérité* (in France), and later observational
documentary (in Britain) (MacDonald and Cousins, 1996). Central to the
new approach was immediacy, intimacy, and "the real". Films in this style
distanced themselves from the polished, professional aesthetics of traditional cinema and accepted images that were grainy and sometimes out of
focus. Despite the new opportunity to approach the real, the film-makers
soon realised that they were faced with new problems and advanced but little in the discovery of "the real".

How was this possible? Direct cinema and *cinéma vérité*, despite similar
intentions, were rather different in how the films were created. *Cinéma vérité*
was based on the view of Russian pioneer Dziga Vertov that the "camera eye"
is more perfect than the human eye in revealing what reality is about. He
provocatively juxtaposed images to create completely new meanings (Ellis,
1979). This way of creating films particularly emphasised the active role of
the author. *Cinéma vérité* was a direct translation from the Russian *kino-prav-
da*, by the French sociologist Edgar Morin and anthropologist Jean Rouch.
Their approach was openly interventive. They used interviews and asked the
people in the film to participate in the process of film-making. For example,
they would ask one of the "actors" to hold the microphone.

Direct cinema in the U.S. opposed this interventive approach. Robert
Drew, who was also a developer of portable film equipment, believed that
with lightweight equipment his film crew was so unobtrusive that they could
record reality without influencing it. Drew, and his followers, focused on

people who were so involved in what they were doing that they apparently forgot the presence of the camera.

Frederic Wiseman, one of the "purists" in direct cinema, however, strongly objected to the entire idea of being able to represent life as it is. When interviewed about the *Titicut Follies* – his first documentary film from 1967 – he described his film as "totally subjective" (Winston, 1995, p. 161). He claimed that:

> *The objective–subjective argument is from my point of view, at least in film terms, a lot of nonsense. The films are my response to a certain experience.*

In the same vein, Bas Raijmakers *et al.* (2006, p. 230), who as designers employ video to create "design documentaries", say:

> *Representations such as film are inherently opinionated because they are inherently incomplete; it is impossible for filmmakers to avoid making choices about what is important. At the same time, filmmakers' biases are constrained by the material they have to film: documentaries cannot simply invent the material they use.*

The question "what is reality?" appears to be an unresolved issue, which no documentarist or scriptwriter can objectively address and settle. So, rather than discuss if designers are able to capture "reality" with video, a more relevant question is how designers employ the video camera in learning about the practice of users, and how this affects the type of material they are able to collect.

Fly on the wall – fly in the eye

How a video camera affects people's behaviour is the topic of ongoing debate. Some researchers claim that the camera quickly blends in with the background (*e.g.* Blomberg *et al.*, 1993; Muller, 1992), while others suggest that one should rather utilise the camera as an active agent to which the observed can relate (*e.g.* Shrum *et al.*, 2004). The debate is largely coloured by the backgrounds and intentions of those who have participated in it. For example, on the side of ethnographers the influence of the video camera on activities seems to fundamentally conflict with the aspirations of the ethnographers – to capture life as it is. The camera and explicit orientation towards it are conceived as biasing the truthfulness of the ethnographic data. (This is not entirely correct of all ethnographers. Some are *very* conscious of their own role in participant observation and how they learn by actively engaging in the situation.) On the other hand, designers employ video to provoke a response in people, whereby their relationship with the tool often seems to be completely the opposite. However, as designers' intentions may also vary from studying what people do at present to understanding the opportunities for changing situations, we need to understand the limits and possibilities of video with regards to both kinds of aspirations.

Brigitte Jordan and Austin Henderson (1995) noted that people's behaviour is influenced by video at various levels. Depending on how automatic or conscious the activities are that people engage in, they may change their behaviour to differing extents. Video may provoke some people to make faces, others to clean up their speech, and yet others to move cautiously in front of the camera. This effect – what the scientists call bias – may wear off as people become familiar with the presence of the camera. Jordan and Henderson (1995) claim that: "Where people are intensely involved in what they are doing, the presence of a camera is likely to fade out of awareness quite rapidly."

Designers, on the other hand, may bring the camera into the explicit focus of activities. For example, Shrum *et al.* (2004) placed the camera in the middle of the table where the users were interviewed. Whenever someone had an idea to share, they would turn the camera towards themselves. Jordan (2000) describes a self-recording method, where the users walk to the camera in a separate location to speak intimately about their ideas and experiences. The video camera turns into the central focus of the activities rather than into a piece of furniture to which nobody pays explicit attention.

The role of video and its influence thus depend on if and how attention is drawn to the camera and video recording. The designer can choose to observe as a proverbial "fly on the wall" or, at the other extreme, to actively encourage people – with the camera as a "fly in the eye" – to reflect on their own practice, and how it might change in light of a proposed technology. However, rather than turn these options into a discussion of right or wrong, a pragmatic attitude must be in place, as Anderson notes (1994, p. 154):

> *This may seem a trivial point to make, but it is not. Once one is aware of it, all the emphasis is thrown onto understanding the processes for patterning observations and their interrelations rather than the methods for recording and summary.*

Even if designers want to use the "unobtrusive camera", their inquiry is always a constructive activity, which seeks to build understanding about a topic. Joris Ivens (1969, p. 228) states that: "The film's art begins when you choose where to place the camera." So, rather than perceiving video recording as data collection, it is more effective to consider case by case how the employed methods will best contribute to the development of relevant understanding and provide resources for exciting inspiration.

The dilemma of relevant focus

User-centred design aims to create products that serve their users. When discussing what needs to be taken into account when designing such products, we are faced with the question of relevance. Roughly said, the users' point of view, and thus design ethnography, is only important to the extent that it is relevant to design. Relevance is a broader topic transcending user studies. Anderson (1994, p. 155) expresses the issue thus:

> *What we will be asking of ethnography is not that it should be a way of getting to know and articulating the user's point of view or whatever, but the analyses it offers us should be directly germane to the interests and issues that confront designers.*

Anderson's statement underlines the importance of analysing the materials constructed in the user studies. It is the analysis and interpretation that renders the material (or parts of it) relevant to design. The following example by Crabtree *et al.* (2002) illustrates the fundamental paradox of relevance. They had the opportunity to utilise over 6000 hours of video material to ground their design of new technologies for domestic environments. The material was captured during a period of over two years. It consisted of recordings from sixteen volunteer households, which had up to five inhabitants each. The cameras captured activities over a period of ten consecutive days in each household. Despite the extensive material, Crabtree later held the opinion that even this abundance of user material was of little help compared to the effort of creating it.

The case is a brilliant example of the fundamental dilemma in conducting user studies for design: the relevance of the material becomes known only afterwards, but the study must be planned in advance! How then can designers ever argue for conducting user studies? In the above example, however, the video material was not captured with designing in mind. Moreover, all the material was captured with a rigid focus. The process lacked the intermediate activities of analysis and interpretation with regards to design intentions, which would have helped guide the study. Thereafter, what the designers needed to do was to browse through a mountain of video in order to discover any interesting themes that could inform design. When Andy Crabtree was asked how he would conduct a study for designing, he confirmed that it should be made iterative.[†]

The key to the solution thus resides in the activity of *iterative framing* of the focus. An open focus makes an enquiry diverse; the sharper a design objective the more focused becomes the user study. During the early phases, the focus is usually open and blurry but clarifies in the course of action through the engagement of various stakeholders in the iterative design events. The focus also becomes partly framed by the project's intentions and possible specifications of earlier models of similar products. Hughes *et al.* (1994, p. 438) also emphasise the value of iteration, which in their study was facilitated by a "quick and dirty" approach and tempered by stakeholder needs:

> *Much of the effort of ethnography was in determining this focus through a series of "quick and dirty" ethnographic studies. An existing focus was also provided by the initial design intentions within the shared object service and the existence of a previous specification within the building society.*

Ethnography as a "thick description" of human culture is an activity that professional anthropologists may spend years writing. Design ethnography is bound to use only a rough version of ethnography, since design projects will not practically allow designers to invest such amounts of time on field studies. Hughes *et al.* (1994, p. 433) again state:

> *The phrase "quick and dirty" does not refer simply to a short period of fieldwork but signals its duration relative to the size of the task. The use of ethnographic study in this category not only seeks relevant information as quickly as possible but accepts at the outset the impossibility of gathering a complete and detailed understanding of the setting at hand.*

Rapid ethnographic research has gained some resistance since it is perceived to produce overly insensitive material, which may cause a design project to move ahead on the basis of an immature understanding, *i.e.* without a proper

† The question was posed to Andy when he was visiting the University of Art and Design Helsinki in autumn 2006.

understanding of the human communities of practice that will be affected by
the designers' work. Acknowledging that designers need to cope with time
pressure, Hughes *et al.* (1994, p. 437) assert that design ethnography essentially provides a means for designers to learn about issues of importance for
designing, also in a rather short time:

> *A charge often levelled at ethnography is that it is a "prolonged activity".
> As we have suggested, this is not quite the problem that it is imagined to
> be. Depending on the purposes of the design, much can be learned from
> relatively short periods of fieldwork.*

The use of interpretation models in contextual design (Beyer and Holtzblatt,
1998) is one solution to the intense time pressure. Here, pre-formulated
schemas for interpretation help designers to focus on relevant issues, especially regarding the design of an information system, and to describe their
findings in an easily communicated way. Moreover, the schemas help to
synthesise findings across a variety of user sites. At the same time, as these
models build on abstracted and pre-designed structures it is likely that they
are insensitive to the flexible ways people actually go about pursuing their
practices. This is where the "quick and dirty" approach may turn out to be
more valuable. As Hughes *et al.* contend (1994, p. 434):

> ... *"quick and dirty" ethnography is capable of providing much valuable
> knowledge of the social organisation of work of a relatively large scale
> work setting in a relatively short space of time. [...]*
>
> *What the "quick and dirty" fieldwork provides is the important broad
> understanding which is capable of sensitizing designers particularly to
> issues which have a bearing on the acceptability and usability of an en-
> visaged system rather than on the specifics of design.*

Ignoring ethnography's value could be much more costly in terms of in-
adequate systems and dissatisfied customers. For this reason, practical
methods have been developed to tackle the time issue in design ethnography. David R. Millen (2000), a research scientist at AT&T Labs Research,
named the approach to cope with a limited time scale in the field "rapid
ethnography". Millen has identified several techniques for quickening the
process, while keeping focused on design-relevant issues. The main ideas
underpin three fundamentals (Millen 2000): study fewer but better chosen

people and activities, use interactive observations, and use collaborative and computerised analysis methods. Along similar lines of thought Werner Sperschneider and Kirsten Bagger, at the User Centred Design Group at Danfoss A/s, have identified several techniques for rapid ethnography with video (Sperschneider and Bagger, 2000). Their techniques – situated interview, simulated use, acting out, shadowing and apprenticeship – intend to move beyond data collection into design-in-context, thus serving tight schedules.

The issue of relevance is two-fold. On the one hand, the materials created during user studies should be relevant for designing. On the other hand, the designs that designers propose should be relevant to the users. Jeanette Blomberg *et al.* (1993) outlined four valuable principles to guide the framing of relevant focus and developing useful materials in design ethnographic studies:

▸ Natural settings: *Studies should be conducted in field settings rather than in laboratory experiments.*
▸ Holism: *Particular actions can be understood only in the everyday context where they occur.*
▸ Descriptive: *The accounts of the human practices describe how people actually behave, rather than how they ought to behave.*
▸ Members' point of view: *The descriptions aim to create an insider's view of the situations and describe the activities in terms that are relevant and meaningful to those who are studied.*

While these principles are very helpful in guiding the design of a project's ethnographic activities, they come short in how they connect to designing itself. Missing from the list is what the art and design documentary authors Raijmakers *et al.* (2006, p. 230) express:

> Design teams may thoroughly research the people and situations for which they are designing, but they must also develop a perspective – a prioritised view – to direct their work.

Participant intervention

Design anthropologists Mette Kjærdsgaard and Gregers Petersen (2007) have coined the term participant intervention to describe a designerly way of engaging with the field through mock-ups and experiments. Their idea stems from their observation on shifting the focus in design anthropology from data collection into a constructive and dialogical process with users. With the advent of design catalysing and mediating devices, such as design probes (Mattelmäki, 2006) and design props, designers may provoke an open-ended dialogue with users. These tools are fundamentally future-oriented, and they act as mediators and placeholders of commonly negotiated meanings. Hughes et al (1994, p. 431) emphasise that designers aim to reconfigure the world that they study, and extensive studies of the current would a waste of resources:

> *Ethnography insists that its inquiries be conducted in a non-disruptive and non-interventionist manner, principles which can be compromised given that much of the motivation for IT is to reorganise work and, as part of this, often seeks to displace labour.*

Sperschneider and Bagger (2000) also ask: "And what about when your goal is not to study social interaction, as in the case of ethnography, but to study change, as in the case of design?" The goal is then the placing of ideas on future technologies (*i.e.* the intended changes) into the practice of people, and then experimenting with changes in the practice and in the design. Design changes the context (including the practices of people), and the context governs what kind of design is appropriate. Hence, designers must find methods that help to discover what it is in current practices that may be changed and how, and what will persist in future. This underlines the need to utilise methods that are able to address current practices as well as to project the possible changes in practices onto the visions of change.

Practices evolve in a discourse with available resources and constraints. When communities are provided with new resources, they may reorganise their practices. These changes are relevant phenomena for a design project, which is likely to trigger such changes. Hence, in order to ensure the creation of good products, these changes need to become the subject of the designers' study.

When designers aim to change situations into preferred ones, they must understand what *needs to be* changed, and what should be maintained. Moreover, they must understand what actually *can be* changed and what will persist. The fact that people's practices evolve through long periods of time enables designers to foresee how things may be in future. Dewey (1910, p. 15) described how artefacts may help to project future issues:

*...things are records of their past, as fossils tell of the prior history of the
earth, and are prophetic of their future, as from the present portions of
heavenly bodies remote eclipses are foretold.*

Hence, the issues can be addressed by designers by entering the sites of
people's everyday activities with the video camera. Through the scrutiny of
materials concerning interaction, the researchers may create so-called "thick
descriptions" of the activity (as we learned from Ryle, 1968), and they may
start to gain a deeper understanding of what forces are at play.

Seeing the activities is, however, not enough. Merely seeing what someone is doing does not relate what affects the work, let alone decide whether
the activity is desirable or not. Is it instructed by someone, or by some rules,
or is it done for sheer pleasure? Martin and Sommerville (2004) emphasise
the relevance of explicit descriptions of a practice as regulating devices:

> *On the one hand it is easy to state that plans and procedures do not cap-
> ture the full details of work or activity as it is played out but the more
> crucial point is to examine the relationship between these and the actual
> "work" undertaken. Where do they (and in what way) guide, constrain,
> and drive action and interaction?*

For developing such a versatile understanding of the studied community
of practice, the use of multiple methods of inquiry may be necessary. For
example, Kjærdsgaard and Petersen (2007) use provocative design tools in
combination with interviews, field studies and other design tools.

Capturing experience

Ethnography focuses on behaviour, but subjective experience is also important. In ethnographic user studies the focus is usually on users' practice in
terms of observable behaviour. Heath and Luff (2000) observed that methods based on ethnomethodology and conversation analysis do not address
the issues of meaning and representation; they are not concerned with cog-
nition and learning; nor do they focus on how the situations shape human
experience and activities. Instead they focus on the "procedural, socially organised, foundations of practical action" (Heath and Luff, 2000).

Designs are, however, in important aspects related to how people experience and make sense of situations. During the late 1990s and at the begin-

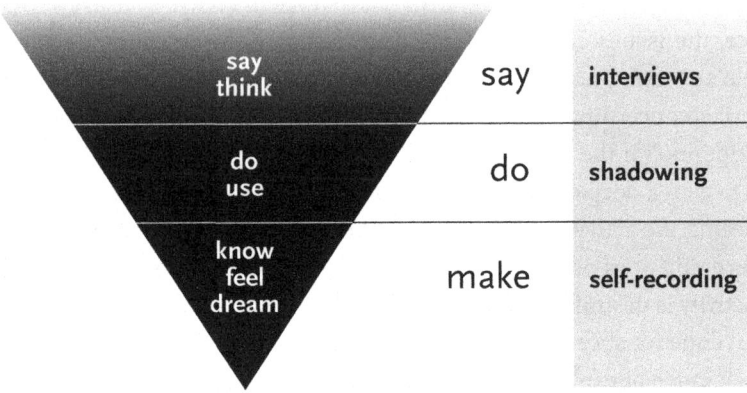

Figure 2.1
Sander's
(1999) "say,
do, make"
framework
and how
video study
methods
relate to it. A
refined model
is presented
in *Visser et al.*
(2005)

ning of the new millennium, emphasis in user-centred design has moved into "user experience". Several academic dissertations have been written on the issue (see, *e.g.* Desmet, 2002, and Battarbee, 2004). Basically the underlining aspiration throughout the user experience literature is the attempt to adopt a phenomenological position in designing, and to include the subjective meanings that are related to products. Such a position promotes the sensuality, meaningfulness and pleasure that are related to the encounters with products.

How is experience addressed in video-based user studies? Liz Sanders and Uday Dandavate (1999), pioneers in developing novel methods for integrating user studies in design, state that: "Each route to experience reveals a different story or picture." Sanders lists three paths in order to access what people know: through what people *say*, what they *do* and what they *make* (Sanders, 2001), see Figure 2.1.

When listening to what people say, a design team may learn about people's conceptualisation of their work or leisure. They may say things that *they want* the design team to hear. Wenger (1998) asserts that in an interview activities may become explained in a way that satisfies the institutional goals of the organisation for whom the individual is working, rather than focusing on describing the real social practice. Furthermore, it is often convenient to explain one's activities on a broad, or abstracted, level that omits a great

expression skills, people will filter their experiences through the expectations concerning the design team's intentions. This provides designers with
explicit material about people's perceptions. However, the picture it creates
is a rather distorted, biased and partial one.

Observing what people do provides a window beyond people's verbal
expression into the *tacit issues* in doing. The following brief example of a
possible study situation outlines how this differs from verbal accounts. In
an interview a technician is asked about his normal routine in the morning at the office. He attempts to convey the details, and he explains as accurately as he can how he browses e-mails quickly, checks the calendar on the
table, and then heads for a client's working site. However, when a design
team goes to observe the activities at the workplace, they may find how the
technician begins the day by talking to a colleague in the lobby, then makes
a quick call to handle reserving some materials, writes a brief note on his
mobile phone about the meeting that the phone call triggers, checks the list
on the wall about the other workers' presence, *etc*. All these details omitted
in the interview may be relevant to the design that will be created later, and
it is precisely these kinds of details of everyday interactions that make up
what practice actually is.

At the deepest level are the issues related to people's thoughts, feelings
and dreams. Sanders and Dandavate (1999) assert that people are able to
express their thoughts, feelings and dreams with tangible and visual tools
that are based on *making*. These "make tools" enable people to express issues on a non-verbal level – yet as concrete ideas. Such concrete descriptions
combined with people's explanations thereof may reveal as yet unknown and
unanticipated, or *latent*, needs and aspirations.

Making, when understood as construction, is a broader topic. The study
procedure in its entirety and the situated construction of new ideas is fundamentally a process of making. It seems that more essential than how one
expresses ("say", "do", "make") is how to build up moments of reflection. It
might be fruitful to understand "depth" as relating to the depth of reflection
both on the users' side and on the interpreters' side.

Entering people's lives

Before designers enter people's lives with the video camera, some issues relating to the risks of videotaping should be considered. Even experienced design ethnographers sometimes have difficulties with real-life organisations,

despite knowing the ethical issues in conduct well. For example, Hughes *et al.* (1994 p. 433) write:

> ...*we may have been unlucky in this case and ... it does highlight an important feature of ethnographic research, namely, its reliance on being accepted in the setting and, even if this is forthcoming, being subject to the range of contingencies that are capable of afflicting all "real world" organisations.*

It is surprising the kinds of damage that can be mediated by the unthoughtful use of video. Physical, mental, social and financial harm are all possible. People are often intimidated by the camera; the video recording reveals their way of being in high detail. Hence, approaching people with a video camera is a highly sensitive issue. The following list outlines some considerations before switching the camera on or even before the first phone call to the users.

Inform the participants about the forthcoming study. This might happen in a phone call. The people being studied can also have the power to affect the timing and target of the study, depending on the context. This helps them to orient to and prepare for the study. It may also help them to think about what they do and what they want to show to the design team.

Attain permission to shoot. Homes are intimate places where everything might not be public. Workplaces may contain confidential plans visible on a table, or people may be present who should not be filmed, such as in hospitals. The space may also feature some tools or arrangements, which form the competitive advantage of the organisation; the filming must therefore have proper authorisation and control by the stakeholders. It is always a benefit to ensure that the design team is authorised to use the video material for the purposes they need. This may require written permission in some cases – and it is a good idea to acquire the permission immediately after shooting, if needed, as the procedure of studying may have helped build a stronger rapport between the parties. (If the edited artefacts are shown somewhere in public, appropriate permissions should be sought so that the users know how they will be presented.)

Be open and sincere. People are expected to express personal details about their lives. Designers need to be open and share details of who they are and what they are aiming at in order to expect others to be willing to do so. When people appreciate the people they collaborate with, and when they feel that they are respected, listened to, and feel that they are able to contribute, it is likely that the design events will succeed.

Explain the procedure. A brief moment of explanation before starting the shooting is usually enough to enable fluent collaboration, as people know what they are expected to do. For example, the user may need to be instructed to work without explaining what he is doing, if the activities will later be discussed in an interview. The user may also be instructed to explicitly point out everything worth noting to the team. Instructing the user to think aloud during the shooting might make sense when there is little time to discuss afterwards. In this phase the user should be reminded to control the shooting: what can and what cannot be captured.

Remind people to avoid physical risks. The presence of the video team may cause people to forget their usual safety routines. Hence, they may need to be asked about the safety issues related to any potentially dangerous interactions. Sometimes people work or have fun in dangerous places. Entering these scenes with a video camera might put the person in danger, which should definitely be avoided. Furthermore, the handling of the video camera might be difficult in such environments, which may endanger the video equipment itself. Thus, to ensure minimal risk to people and to the equipment, the design team needs to inform the study participants about the possible physical risks in the study and give instructions on how to avoid them, and *vice versa.*

Inform others. In shadowing studies the people being studied quite often meet other people during the video recording. When possible, it is helpful to have the others informed about the study in advance. In our procedures, we ask the person being studied to briefly explain to others the purpose of the research; how thorough the explanation needs to be depends on the person encountered. Outsiders may be edited out of the footage if they happen to be visible in the video recording.

Avoid making a fool out of anyone. Editing can turn the same person on the video into a bright-minded thinker, or an ignorant troublemaker. It is often a matter of choosing certain clips and placing them in a specific order that creates this meaning. People are precious collaborators and must be considered with care.

The above list of ethical principles applies to all video-based activities throughout this book and is helpful in avoiding major problems in a design project. When designers are fully aware of these issues, they may move ahead to study the constructive co-authoring of design-oriented video materials, which are explained next.

● Method: Situated Interview

*"Could you
explain
what that
is for?"*

Interviewing is a widespread method in social studies to explore what people think about things. "Being situated" means having direct access to the details of the practice within the moment of the interview. This may mean conducting the interview in the usual environment, such as at the work desk of the user, or bringing images or tools of the worker to the interview. This allows a more detailed discussion on the particular relationship between the person and the issues in focus.

Interviewing is fundamentally about someone asking questions and someone else answering them. However, the configuration may vary from intimate and deep individual reflections to group interviews. The situated interview is focused on studying the "real" person in the "real" setting. Hence it differs from the kind of interviews conducted to build an overview picture of a larger whole. Thse may be carried out, for example, in interviews with the workers' superiors.

Practical guidelines
- Start with easy questions.
- Prime the interview with self-documentation, or use observation as a help to being more reflective.
- Ask open questions rather than brief "yes" or "no" questions.
- Provoke details through details: Ask concrete questions and provide a detailed context.
- Get a real practitioner: Remember that someone who thinks she/he knows, such as the superior, does not have the same relationship to the practice.
- Ensure good sound quality: Use an extra shotgun (or wireless) microphone in noisy environments.

An interview is useful when a design team wants to edit video portraits of people. A personally expressed spoken story conveys the meanings the ma-

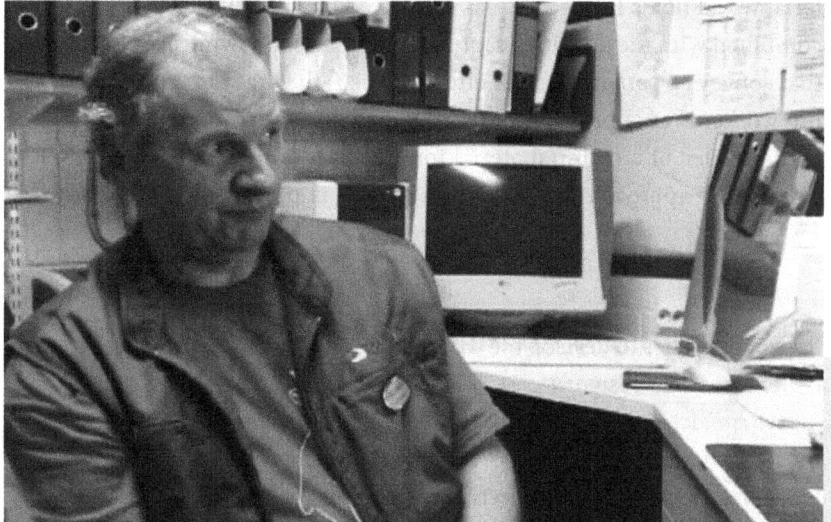

▶ Case story: Ageing workers

Salu Ylirisku and Kirsikka Vaajakallio, University of Art and Design Helsinki

Video
example
**Interview at a
schoolhouse**
2'58"

The schoolhouse caretaker is sitting in front of us at his work desk, his shirt sweaty after working intensely for one-and-a-half hours. Salu has placed the video camera on a tripod and is holding a sheet of paper containing roughly-structured questions. Kirsikka is preparing the laptop computer on a nearby table for the display of the still pictures captured during the shadowing done just before the interview. We aim to create a soundtrack for a user portrait that we may edit using the worker's comments on the situations presented in the still pictures. Before starting the interview Salu checks that the external shotgun microphone attached to the camera is on. We are quite excited, as this is our first interview as a team in this project.

The Konkari project was part of a two-year EU-funded research project (2004–2006) to improve the well-being of ageing workers. The project was conducted at the University of Art and Design Helsinki, and the ageing workers were employed by Palmia, a company owned by the city of Helsinki. Palmia provides catering, security, cleaning, and technical maintenance services. Our study focused on the latter two of these. The participating ageing workers as the focus of the study were all over the age of 50. The workers' interviews were conducted to study the workers' thoughts about their work, terial world holds for people – as conceived by them. When such a story is combined with the activities captured during a shadowing event, a portrait that conveys a person's values effectively is rather easy to create. The "Ageing workers" case provides an example of a situated interview conducted with the idea of creating material for video portraits in mind. A completely different approach to situated interviews is presented in the "Freeride

the opportunities to develop the work and also to construct engaging video material to drive design.

We contacted the workers some two weeks before the site visit. We asked the worker to choose the time for the study, and said we would be observing the real work practice. The observations and interviews were conducted in schools during the daytime when the pupils and teachers were present. We thus also needed to attain permission from the headmaster of the school for the study.

When we arrived at the work site, we first met the worker and briefly explained the idea of the day: first we would shadow one and a half hours of continuous work, after which we would conduct a half-hour interview. We also explained that we would be like proverbial "flies-on-the-wall" during shadowing, and that we had the chance for discussion afterwards.

After the shadowing was over we moved to the interview. We had a four-point structure: (1) the person's background, (2) today's activities, (3) future opportunities, and (4) the personal message for future colleagues. The observation phase combined with earlier activities in the project had familiarised us with each other quite well. It was thus not particularly difficult for the participants to give a relaxed interview. We thought this would be helpful in the construction of the video portraits. The overall aim of the project was the well-being of the ageing workers, and this was seen to be influenced by the ways people understand their role in the organisation. The user portraits that we aimed to create underlined the value of the ageing workers.

The interview questions combined with the still photograph "playback" of the situations provoked brilliant material for the later editing of the videos. Moreover, the observation session combined with the interview enabled us to gain access to the real-life interaction as well as the workers' thoughts about the work. ■

▶ Case story: Freeride skiers

Salu Ylirisku, University of Art and Design Helsinki

Video examples

Show your stuff interview *1'34"*

About to ascend *2'14"*

Distant shadowing *3'28"*

On the mountain *2'06"*

A cold sea breeze from the Arctic Ocean blows dark clouds above the horizon from behind a smooth ridge, where six freeride skiers together with two members of our research team are hiking in May 2003. The camera will survive the snowfall, though, since I have some plastic bags and sticky tape with me to protect it from getting wet.

I am a bit worried about the weather getting worse, since I may not be able to see the skiers, and I may get nothing but a white curtain of snow on the videotape. I am standing in a pit that I have dug to protect myself against the cold wind while waiting for the skiing to start. I keep the extra batteries for the camera in my pockets close to my skin to keep them warm. The snow is hard up there, so it was relatively easy to get the tripod to stand firmly on it. I am wearing woollen gloves with open fingertips under thick leather gloves, which I will remove when videotaping.

Finally the skiers appear from behind the peak far above me. "Salu, do you read?" I hear from the radio. "Yes, I do," I reply. "We'll start from here with Jani. Tell us when you are ready with the camera," says Antti, who is a member of our research team. The others continue further up on a steep crag. Despite zooming in as close as possible, the frozen and slow LCD screen displays the skiers as tiny black spots on the texture of the mountainside. I wonder if it makes any sense to videotape these dots.

skiers" case, where video was employed in various situations in a rather exploratory manner. These events, however, provided useful material for authoring video artefacts about freeride skiers' attitudes, as presented in Chapter 3.

Interviews are most useful for design projects when they are utilised to complement other methods, such as observation and participation in the exploration of users' reality. Interviews may be conducted with provocative materials that help to orient thinking towards design opportunities. Such an approach is presented in the case "Ageing future" later in this chapter. In a sense, such an interview is situated in the context of ideas about the future. ∎

The Luotain project (2002–2006) aimed to develop user-centred processes for product concept design with an emphasis on user experience. The project, which was mainly funded by the Finnish Technology Agency TEKES, included in total seven case studies exploring particular methods and tools to capture and represent user experience for design. The freeride skiing case was one of these. It aimed to construct an image of freeride skiing sport equipment for the Suunto Corporation in order to help design interactive sports instruments for skiers.

The process included expert interviews, a literature study, and a probes self-documentation period of two weeks with six freeride skiers before we went to observe the actual skiing with six skiers on the Lyngen fjord in north Norway. We lived for four days in the skiers' hut and during this time we had plenty of opportunities to videotape the activities. However, we found that the videotaping of the informal conversations was a bit problematic. We wanted to maintain a casual and informal atmosphere, but the camera in our hand tended to turn the discussions into interrogations rather than lively debates. Hence we adopted a strategy to leave the camera aside for the chatting and instead wrote notes after discussion. Our research team, which consisted of me (the design researcher) and three Suunto personnel (one product manager, one concept designer and one usability specialist), were able to discuss the findings and reorient the focus when driving to the skiing locations. Some of these we captured with video as records of the key findings.

The rather long period with the skiers allowed us to try out different ways of capturing the activities on video. In the hut we had several organised interviews,

Shadowing is a method for observing people while they move. The metaphor of shadowing originates from detective stories. Like detectives, the designers with the video camera try to build a record of what a person does, where she goes, which equipment she utilises, and who she encounters. Unlike the subjects of detectives, the studied people know well who are observing them and for what purposes. This allows close cooperation in building material that is valuable for design.

As mentioned earlier, many work activities are automatic and are thus difficult to verbalise or to detail, or may even escape conscious awareness

"May I follow you to see what you do?"

where we had some prepared questions based on the findings from the previous phases. A couple of the skiers were present in these. We had the chance to observe how the skiers prepared for a hiking trip, how they planned where to go, how they packed their bags, what they ate, how they observed the weather, *etc.* During the skiing we had three cameras running in parallel. Two of the cameras were held by the research team, and one of the cameras was lent to the leading skier with the instruction to record and think aloud what he was thinking in various spots on the mountain. This worked surprisingly well in this case, perhaps because the skier had some background in videotaping. When we watched the video recordings in the evening together we also had the chance to hear the skiers' comments on the day's activities.

One of the most interesting bits of video material that we captured was a situation that might be called the "show your stuff interview". One of the skiers spread out all the skiing equipment on a blanket and he explained the purpose of each piece of equipment while I was recording the interview. It provided us with a condensed information package on how the skiers think their equipment relates to their activities.

The case study provided us with extraordinary video material with highly engaging content. Despite not having a fixed idea of what to shoot during the trip, the presence of the camera allowed us to discover new uses while we were there in the field. Based on this experience it seems important just to have the camera available. Utilised with an exploratory mind it may prove to be quite useful. ∎

altogether. Shadowing produces material on the details of everyday interactions in people's usual environment. When shadowed (and usually all the time) people tend to make their acts intelligible and somewhat predictable in advance through hints, such as orienting towards something, nodding and pointing with their eyes. A designer who follows these clues is able to move the camera according to the focus of the subject and build a video that becomes a sensitive rendition of a person's characteristic way to go about things.

In shadowing the signalling of intentions is a two-way activity. With the ability to control where the camera is pointed, the designer constantly signals users as to the areas that are interesting for design. This often provokes users to show things to the designers with the video camera. Hence shadowing is a method that calls for sensitivity, quick response, skill in reading the subjects' focus of attention, and the ability to inspire collaborative exploration in order to orchestrate the interactions towards a design-driving result.

Some practical issues when shadowing
- Keep the person in the picture at all times.
- Follow what the user is doing and where his/her attention moves.
- Use your feet to zoom.
- Keep up with the pace of the user.
- Remember that if you cannot hear, neither can the camera.
- Let the video run continuously (do not stop the camera when surprised).
- Allow the "user" to control what can be videotaped.

In multi-camera shadowing, a design team approaches the user site with several video cameras. This makes sense in cases where several users are interacting with each other across a distance. Such cases may occur when a working group consists of several people whose physical areas of work are separate. Multiple cameras were utilised in the "Plant operators" case, which focused on exploring the way wastewater treatment is conducted by the operators of the process. With such video material designers may edit stories that convey how the procedure unfolds with multiple persons involved. These multi-camera videos provide a "God's eye view" on the interactions, which no single person is normally able to achieve. ∎

▶ Case story: Plant operators

Jacob Buur, Danfoss User Centred Design

Video
example
**Plant
operators**
2'30"

Monday morning often means trouble. It is Monday morning at the Himmark wastewater treatment plant. Flemming, the lab technician, is going
about his daily routine in the small chemical lab. He is analysing samples
taken this morning from various basins of the plant, to check the level of
pollution. Christina and I have been permitted to follow Flemming's work
with our video camera for one day. Christina is a PhD student from Aarhus
University, and I work with the Danfoss User Centred Design group. Right
now we are with Flemming at a bench with lab equipment, I with a handheld
camera, and Christina next to me, trying to find a balance between when to
ask questions and when not to interrupt the work.

Flemming has been animatedly describing in detail why and how he
analyses the samples in the small glass caskets, but suddenly he is very
still – one of the glasses has taken on a dark blue colour, much darker than

the other samples. Flemming gets up, and strides quickly out the door. Should we follow suit with the camera? Or wait here? Is he simply going to the toilet? We decide to chase after him as he calls down the hall for the head plant operator:

– Ole?

As Flemming rushes to meet Ole in the corridor we suddenly find ourselves facing Kirsten and Ingrid, who have their video camera pointed at us. They are part of our design team, shadowing Ole. What now, should we turn off one of the cameras, to preserve tape? Better not.

– It's all wrong out at Holm, says Flemming to Ole.
– Really? How high is the level?
– It's above 7 at least, more than I can measure.
– Well, I'd better go out and check, then.

Flemming returns to the lab, while Ole prepares to drive the five kilometres to Holm, an unmanned satellite plant. This little incident starts a string of events, much like the Three Mile Island disaster, only much smaller in scale, of course. And we happen to be there with three video cameras running!

The water vision project. The wastewater treatment plant field study was part of a vision project on new technology for the water business segment, organised by the corporate User-Centred Design group of Danfoss, a major Danish manufacturer of industrial controllers. Danfoss has several business divisions that develop products for wastewater plants: pump controllers, flow meters, pollution sensors, automated valves, *etc*. The goal of the project was to study the water treatment field from a user's perspective and suggest a vision for Danfoss products and user interfaces. As in many other industrial plants, the situation for operators is changing rapidly, with more and more computer control being embedded in the products, and products being linked in networks.

In the project team we were ten in all: user-centred design specialists, developers from business units, management trainees, and university students. In total the project took ten months with two months spent on user studies. It was organised in collaboration with two other research teams from the Universities of Aarhus and Malmö, which allowed comparative field studies at three wastewater plants.

To study the people who work at wastewater treatment plants posed quite a challenge. Plants are large installations with walking distances of up to several kilometres. They are manned by a staff of eight to ten operators, who work with mechanical, chemical and biological processes, which were all new to the team.

Our initial contact at the local wastewater plant was with the head plant operator, Ole. At our first visit (two of us), he kindly explained the good a plant does and how it works. It must have been all too obvious that our engineering and HCI training had not prepared us for understanding wastewater treatment at any professional level, for Ole comfortably switched into his school children routine, explaining everything in simple, pedagogic terms. He had a map ready, showing the complicated flow of water and sludge, and even a little pamphlet that listed who works at the plant, and what they do. Then he took us on a tour of the facilities, in what we later found out was his daily morning routine. We noticed the walking distances, the smells, the machinery, the abundance of chemical terms, and also the subtle cues Ole apparently took notice of. We were kindly allowed to videotape the tour, and thus had material to show the rest of the team.

The first video recording started quite a discussion with the team and colleagues in Aarhus and Malmö about how we should go about the user studies. How much time should we spend? How many of us should go? (We all wanted to!) Where should we start?

We badly wanted to observe work revolving around Danfoss products, but to stand and wait at any one product for something to happen was clearly not a workable strategy, as they are not operated on a daily basis. We decided to use an ethnographical approach, studying the activities of several operators as they unfolded simultaneously. Based on the overview of employees, we asked permission to shadow three employees for a full day. As we had heard that Monday was often the most stressful day (after a long, unmanned weekend), we specifically made the appointment for a Monday. Similar appointments were made in Aarhus and Malmö for days within the same week, and we decided on a rotation scheme, so that someone from the two other teams would always join a local study.

Shadowing three operators simultaneously. At 6 am that Monday morning in late September the team assembled in the parking lot outside the plant. The six of us divided into pairs, each ready to video shadow our operator. We synchronised the camera clocks to make later analysis easier and checked batteries and tapes one last time before entering the plant. The three operators welcomed us,

One operator analyses the water

had a little laugh about their future careers as Hollywood stars, then set out to start their work in their respective areas of the plant.

Ole, the plant operator, started his day with a plant walk-through, checking on all the running processes. He used his eyes, ears, hands, and nose to sense any abnormalities in the plant operation. Then he was called upon for a variety of activities through the day, and finally sat down at his desk to complete administrative tasks.

Flemming, the technician, first took samples at several locations in the plant, then spent all morning analysing them in the chemistry lab. He also performed tasks related to the computer monitoring system.

John, the electrician, started his day working on a new pump controller installation (with a Danfoss product), then was called to fix a problem elsewhere. He also had routine maintenance on his agenda.

One lesson we quickly learned when video shadowing is: Never stop the camera recording! For one thing, it is difficult to synchronize three cameras later, if there are gaps in the recordings. More importantly, one cannot anticipate what events will come and which ones will be important for the study. In the lab, for instance, if we had stopped the camera, we would not have been able to trace back what actually happened, or which event led to which.

When two shadows meet. With multiple cameras following people, surprising instances may occur. Sometimes, when two operators – with their shadows – met for a brief talk, we suddenly found ourselves videotaping another crew who was videotaping us. In this way we also learned how well-developed the operators' sense of each other's presence is. At one point, for instance, Ole leaves his office and walks to the top of an outdoor staircase to shoot a question at

John, who moments later happens to pass by at the bottom of the stairs. How could he know? Another instance that happened a few times while we were at the plant was that one operator would call another on the phone for a short discussion – and we would have a camera at each end of the line! Ole, for instance, when arriving at Holm and finding a polluted basin, calls Flemming back at Himmark, asks him to log into the control system, and guides him to shut down a pump station to prevent more wastewater being pumped into Holm, while they investigate what is wrong.

The Holm breakdown. That particular Monday proved to be just as stressful as we had been warned – or even more so. Flemming's lab sample turned black, and Ole was alerted right away: an unmanned satellite plant (located at Holm) had an unacceptably high pollution level. This required immediate action, so a series of events unfolded over the next couple of hours, involving problem diagnosis, replacement of a defective dosage pump, repair of a short-circuited power line, and a report to the local environment authorities. Incidentally, it even involved a problem with a Danfoss component.

To reconstruct the course of events took a good deal of hard work, because it involved activities covered by all three cameras. One might say that we were awarded a kind of "God's eye", a perspective on the events more complete than any of the involved themselves would ever be able to have. Just like in a directed theatrical movie, we were able to cross-edit the activities of three people to show a more interesting story.

Feeling, watching, controlling. One may assume that unmanned plants are the key to rationalising wastewater treatment in the future. For a number of reasons, we learned that this is not the case. As a result of the user studies, we found three keywords that nicely summarise the work at wastewater plants: operators feel the state of the subtle processes using all their senses, not just computer displays. They watch the industrial components, because they know from experience that they are potentially unstable. They control the control system, because automatic systems are really designed for "normal" operation. When special conditions eventuate, a human has to take over with the experience of years of work. Based on the understanding achieved through video shadowing, we were able to generate ideas to support operators in those tasks. ∎

In-situ acting is a method for studying people's practices in their native settings. *In-situ* acting developed partly as a response to the difficulty in observing real activities of real users in their real setting within the tight schedules of design projects. Another reason for its development was to overcome the barrier of the differing professional languages of users and designers. However, perhaps the most important reason to employ *in-situ* acting is the flexibility that it allows for designers and users to explore and experiment with situations (both current and potential) that are considered relevant to the design project in question.

Fundamental to user-centred design is placing designs into the context of use and evaluating the value of the ideas there. *In-situ* acting aims to construct the context as accurately as possible in order to ground exploration and possible ideation to the details of real practices. It uses the same presentation format in which the practices exist in everyday life, which enables interpretations to be built on records unfiltered by the abstractions of language. Even though acting out does not directly correspond to real activities, it does provide opportunities for learning about the details of the users' practice – details that would remain silent unless provoked. Moreover, the delightful atmosphere that the idea of acting out instigates is helpful when people explore radically new opportunities.

Acting out is also employed in the realm of documentary film-making to co-create detailed illustrations with people about their practices in their respective cultures (Raijmakers *et al.*, 2006):

> ...[C]o-operation ... makes people participate in the film differently;
> they are more involved. Building on participation and co-operation,
> Rouch [as a key example] pushed the boundaries of cinema and
> anthropology resulting in what he calls "ethno-fiction", fusing description
> and imagination in anthropology, and realism and fantasy in film.
> Chronique d'un Été contains several scenes where a protagonist is role-
> playing and being herself at the same time. The point is not whether she
> is acting or being herself. The point is that it is not relevant one way or
> the other: in everyday life, "role-playing" and "being oneself" co-exist,
> and the relationship between them is more important than either one
> of them.

- ▸ Frame the situation in a proper environment with appropriate tools.
- ▸ Prepare props if future-oriented acting is desired.
- ▸ Establish a relevant orientation: When, who, and what are usually good facilitating questions.
- ▸ Use video in the same way as in shadowing.

Thank you, as soon as possible.

▸ Case story: Ageing future

Salu Ylirisku and Kirsikka Vaajakallio, University of Art and Design Helsinki

"When should we propose that he could use the camera functionality for this?" we ponder as we are capturing schoolhouse caretaker Seppo in action. Seppo is acting out a situation where he uses the mock-up product that he has designed for his work.

– "I do not know the exact model, but it is one of the round-shaped Arabia sinks," Seppo replies to the imaginary service attendant on the phone.

Video examples

Thinking bubble 2'01"

Tool reflections 2'21"

Reporting a toilet problem 1'37"

When acting is organised at users' sites, the users tend to feel rather comfortable, compared to being invited to a design studio to act. Acting as oneself, moreover, does not entail the trouble of pretending to be someone else, which is the realm of professional actors.

The case "Ageing future" shows how the *in-situ* acting approach facilitates an open and flexible study of the potential change into the user practice. During the project video material was created both of the workers' activities as they would normally occur and of situations acted out by them. In addition,

We decide to remain silent and continue capturing how Seppo goes on with the situation. We believe that by delaying our question about the camera functionality it will help us discover something new – perhaps a nice workaround to the situation.

– "It is here on Albertinkatu (Albert Street), fourth floor, girls' toilet," Seppo explains, holding the mock-up close to his mouth.

So, the location was the next thing to communicate. He then presses a button to store the event in the memory, and then another button to transliterate the discussion into text for the automatic generation of an order form. At the same time, he continues to explain sarcastically how the form would automatically be sent to a city bureau, but as the bureau is a bit behind in technology, he would need to print the form and send it by mail.

Only now that Seppo has finished the action and is leaving the toilet do we propose the camera for communication. "Yes, you could do that. That just did not occur to me since I so seldom send photos."

The Konkari project (which is explained in the case "Ageing workers") also included a phase where the workers' practices were studied and design opportunities were explored with an interventive approach. We called the approach "situated make tools" (Ylirisku and Vaajakallio, 2007), and it takes Sanders' idea of make tools to the real activities of the workers. The situated study was conducted with 12 workers in total. In the first six studies we utilised only shadowing, and in the six subsequent studies we asked the participants to create a tool with the make tools kit that would help them feel better at work or to work more focused.

The study had four main aims:

the discussions with the workers where ideas were evaluated were filmed.
The case illustrates what wonderful actors workers may be, and that the observations may greatly benefit from the imagination of the workers. When
designers are looking at the practice with an "eye to change" rather than with
an "eye to observe", they begin to form numerous ideas themselves – and validate these in the real setting with the user. Combined with the wealth of ideas
from the users themselves, these may provide designers with an invaluable
resource in the later phases, as happened in the "Ageing future" case. ∎

1   to create concrete and relevant-to-the-worker design ideas expressed in
    physical, narrative and acted-out formats;
2   to develop insights into the workers' needs, desires and attitudes relating
    to digital information and communication technologies (ICTs);
3   to explore how the real-action context triggers and grounds inspiration for
    concept design;
4   to gain experience in how the make tools function when used in the midst
    of everyday activities with ageing workers.

The study began by contacting the participants. They were asked to bring a digital tool that they use every day at work to the event. Interactions at the workers'
site began with reflections on their tool: where they would normally utilise their
own digital tool, how they use it, and the kinds of situations where they had
previously used it. The exercise aimed to provoke thinking towards the potential of new ICTs. This discussion and reflection lasted for a half-hour. We then
introduced the make tools kit.

After hearing our instructions the workers started to figure out possible
shapes that would suit them. We also gave the worker an additional instruction
to explain the purpose of each piece that was included in the tool. We asked the
worker to relate the purpose of each new feature in relation to a specific situation. We repeatedly asked the worker to think of existing situations and tasks
where the tool might be helpful. We proceeded very slowly during this phase,
to allow the worker to take the time needed to think about the work from this
given perspective. Here, we considered it very important to enable the worker
to relate the design to the real-life situations and to the needs in these situations in order to ensure the ideas' relevance.

Self-recording is videotaping done by the users about their own practice. It is a method that allows the users themselves to decide what to capture, when, where, and how. It enables them to construct stories and material for further exploration by a design team. Self-recording may focus on documenting interactions with existing practices, capturing an individual's thoughts, or propelling the making of visual stories about experiences with products – both current and potential.

Before starting the shadowing we instructed the worker to carry on with work as usual for a period of one and a half hours. We explained that we would be shadowing with a video camera, continuously recording the activity like proverbial "flies on the wall". And, occasionally we would interrupt the work, if we perceived potential for using the tool that the worker had designed. We called this intervention the "thinking bubble". This moment was geared to discussing how the tool could be utilised in the activity and to envision how the situation could be changed with the tool. Then we began the observation.

Evaluation of the ideas immediately challenged the designers' conceptions of what is needed. For example, in one situation the worker did not accept the idea of camera-based communication for the task of repairing a water tap. It seemed evident to us that the worker would need to communicate to a plumber through images of which tools and parts were needed for a certain tap. However, the worker objected, since he had been with a plumber so many times previously, dealing with the chemistry school's special taps, and he had needed to explain the mechanisms by physically instructing the plumbers how they functioned and which parts needed fixing. This was a surprise to the designers and helped to refine the ideas.

Ideas on site. The situated make tools approach provided us with many design ideas already at the user site. This differed drastically from the previous approach that utilised only observation. We think that the orientation of the designer towards the site is considerably different when the approach is interventive compared to when it is not. We believe that the interventive approach helps one to see the situations with a designerly "eye to change" compared to the "eye to explore" that is active during observations. ■

Since the first experiments with *Cultural Probes* – provocative self-reporting kits to involve users in design projects (Gaver *et al.*, 1999), self-reporting
has established its place in the set of methods that user-centred designers
may employ in their practice (Mattelmäki, 2006). Self-recording is a constructive activity (like using a probes kit) where the users build images of
the issues outlined by a design project team. Raijmakers *et al.* (2006) talk
about self-recording in the form of video diaries:

> *Video diaries are useful for user studies because they can give access to*
> *people's everyday life on a very intimate level. The dialectic between the*
> *maker and the situations she/he talks about still exists in video diaries,*
> *however. Makers of video diaries in fact perform that dialectic in front*
> *of the camera when they reflect on things they did or situations they en-*
> *countered, since they choose what to present and may overlook taken-for-*
> *granted details of their lives. The video-diary is a good way to learn what*
> *people think; it may complement methods such as ethnographic observa-*
> *tion that can reveal what people do.*

Self-recording is helpful in studying processes that unfold over a long time
period, such as a week or two. It allows designers to address situations in
intimate places, like homes, without being there and disturbing the activities.
Self-recording may also be practical in places that are too hard to access for
the designers, such as in the "Freeride skiers" case. Self-reported material
usually requires an interview to discuss the meanings that the users try to
convey through the materials they have constructed. The material may as
well be utilised as such to inspire design.

Guidelines for self-recording
- Instruct the person on the use of the camera.
- Provide a focus: describe the kinds of issues the project is interested in.
- Explain how to deal with other people that may be videotaped:
  Hand out, for instance, a brief outline of the project that helps the user
  to explain the project to outsiders easily.
- Inform the user how the videotapes may be utilised later.

The following case story "Lemmu the cushion" illustrates the importance
of providing the users with proper instructions, and shows how events may
not go quite as the designers expect despite instructions. The Lemmu case

also explores the issue of documenting the experiences that users construct in their own settings. This is also in focus in the case "Mobile experiences", which illustrates how important it may be for a design project to allow users to document their own experiences with new products. In this case it reveals how designers discovered the influence of the presence of the researcher on the ways people try out new applications.

The "Freeride skiers" described an example of self-recording in a place that was not accessible to the researcher. The mountain was simply too dangerous for a novice climber to attempt; one of the skiers himself thus documented activities during a hiking trip. The thinking aloud of the leading skier, who was filming, helped designers grasp what the skier was thinking in different spots: how, for example, he chose the route to the mountain top, what he thought about the snow conditions; and how he saw the team of skiers around him. ∎

Co-exploring

A concept that conveniently summarises this chapter is that of *co-exploring*. Conducting a field study for a design project is much more than trying to capture the objective data of an undisturbed reality. For the designers, it means entering new realms of user contexts and practices, and designerly interventions may help to understand both what is there and what may change in the future. Exploring is a means to encounter the new – whether surprising or expected.

For the users, the reflective process that is triggered by the very presence of designers – and even more by their questions and suggestions for future technology – may enable them to see their practices in a completely new light. Exploring may mean an increased awareness that already in itself instigates a change in the practice of users.

Co-exploring is a particular view on field studies that helps us see the study not simply as questions asked and answers given, but as a participatory endeavour, banking on the combined efforts of users and designers to move towards a better future. The video camera is a convenient "excuse" to set this process in motion: a tool for which and with which we may explore.

▶ Case story: Lemmu the cushion

Katja Battarbee, University of Art and Design Helsinki
Anne Soronen, University of Tampere

The video opens with a living room scene and two small girls exploring the contents of a plastic bag that they are holding between them. Their blond heads are together and they are about two and four years old. A large, fuzzy, cowhide patterned cushion is on the floor behind the girls. The four-year-old girl gets hold of a small object in the bag, walks over to the cushion and presses the object to the cushion. The cushion emits a sudden growl-like sound, and the girl jumps up, shrieking with laughter and dances on tiptoe back to where the bag is to get something else to try.

We laugh, too, at their excitement and our own relief at having data. Many questions come to our minds as we watch the family members explore, struggle, smile and cuddle with the prototype. Scribbling notes, we pack the minidisk and set off to meet the family in person.

The Morphome project investigated issues around designing proactive technologies for the home environment. It started in 2003 as a three-year cooperation project between the University of Art and Design Helsinki, Tampere University of Technology and the University of Tampere, funded by the Academy of Finland. Lemmu, a cushion prototype used in the study, contained an RFID reader in a padded pouch. It was built to demonstrate that sophisticated technologies may look non-technical and cuddly on the outside. When an RF tag was laid on the cushion, the cushion emitted a short sound: a whistle, a chirp or a roar – depending on the tag. The prototype aimed to help explore how technology-mediated everyday experiences become constructed in homes and provoke thinking towards future opportunities.

Three Finnish families were recruited during autumn 2003 to take part in a week-long evaluation of the prototype in the home. We wanted both real footage on video as well as interviews and discussion, and chose to give the digital video camera to the family so that they could document their experimentation themselves. In each of the homes one parent took charge of the camera and prepared to document the situation as the children were given the cushion and the tags to explore.

After the first home we decided to create a small booklet with simple guidelines describing the kinds of things that we as researchers were interested in. We asked the people to avoid overdone staging and propping, and we emphasised just letting the camera roll. The booklet was helpful in sensitising people to think about the issues before the interview.

The families dispatched the video material to us after the study period was over. We looked through the videos once before the interviews that focused on how both parents and children conceptualised their use experiences. The interviews were conducted in the homes, where the family members could show us things and places that were not always clear in the video.

Video was helpful in the study of the responses that the cushion provoked, and the interactions with it. Some of these reactions were easily perceptible on the video, such as the brutal treatment the prototype received – which was actually quite a shock to one engineer in our group. Some interactions needed more work on the material, and the most interesting ones were inspected in detail. In one of the homes, where under-school-aged girls experimented with Lemmu, the transcription of the first 90 seconds of their exploration revealed a systematic, iterative testing of hypotheses on the functionality of the cushion (see Figure 2.1).

Some video stills were rendered by hand into line drawings due to privacy and permission issues to enable communication of the findings. This proved to be a surprisingly useful act. The advantage of drawings over small video still images is that the line drawings can be easily reproduced with black and white printers, also in smaller sizes; the stills were often fuzzy and would not have reproduced well. In the drawings it was also easy to bring forth relevant details and leave the rest out of the drawing. This technique provided a quick workaround for several issues at the same time: resolution, image quality and privacy. The making of the drawings also made us study the interactions, body language and positions of the children very carefully – helping to see details that would not have been perceived in a single viewing.

Creative response to the instructions. The parents followed the instructions at least during the first day of the study, when they dutifully recorded their children figuring out the prototype. The adults interpreted the cushion primarily as a children's toy, which is a likely reason for their slight unwillingness to interact with it when the video camera was recording. Our choice to give the camera to the

Figure 2.1
Girls testing
the Lemmu
cushion

Jutta leans back and lets Riina press the blue tag into the cushion.
— *(Lemmu)* DOOLEE!
Riina spins around to pick up a second tag into her right hand.
— *(Riina) now this*
Riina presses both tags in her right hand to the cushion.
— *(Lemmu)* DOODAA!

[Hypothesis: maybe if you use two tags together the sound will again be different. Result: possible, but it may just be the new tag as well.]

families and request videotaping of particular kinds of situations gave rather free hands to the participants. This meant that many situations that would have been interesting for us were not recorded for some reason or another: they forgot; the situation was over too soon; or they did not want to bother visitors by asking permission to record, or any other such reason.

The idealised image. The video material did not merely document all what happened, but provided constructs that promoted a certain image of the family. The parent operating the camera decided which room or viewing angle to use, who to record, when to start and finish recording, *etc.* Regardless of the research method it seems that the informants want to produce a particular kind of, and often idealised, image of themselves as users of technology. ∎

▶ Case story: Mobile experiences
Minna Isomursu, University of Oulu

On a sunny and busy weekend in downtown Oulu, we give pairs of users two devices – one with the application to be evaluated and the other, a mobile phone with video recording capacity. When we watch the video clips on the following Monday, we are surprised. The clips reveal to us a completely new perspective on the use of the application. The emotional responses, especially, are radically amplified when captured by the users themselves. These expressions help us to identify the lurking design opportunities beyond the other material we already have.

The Rotuaari project aimed to evaluate context-aware mobile applications in a real-world environment with real users. The presented case took place between 2001 and 2003 in Oulu, in northern Finland. The context-aware applications evaluated were a location-aware map and a context-sensitive advertisement. The study utilised a technique called "experience clip" (Isomursu, Kuutti and Väinämö, 2004): a pair of users were given two mobile devices, the application device to one, and the video capturing phone to the other. The instructions to the observer were the following:

*2 Studying
what
people do*

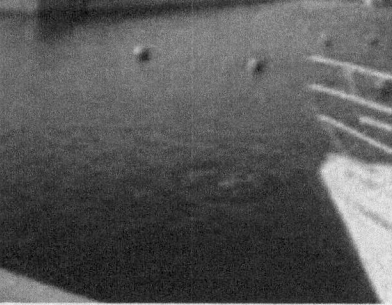

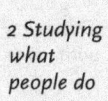

Video
examples

**Dislocated
cultural
centre**
0'08"

**New
interactions**
0'08"

Scenario play
0'10"

**Visioning new
features**
0'07"

– Record as many clips as possible.

– Focus on use experiences: failures, success, surprise, joy, anger, *etc.*

– Aim at the user of the PDA, not at the PDA screen.

The material was captured during three weekends of the one-month field experiment period. During the experiment, a total number of 36 people acted as observers with camera phones.

Towards natural use. The influence of the presence of the researchers became clear when we compared the new experiences with an earlier shadowing study.

When the researcher was present, the users did not try out anything outside the scope of our research or intended usage of the device. They also tried to avoid situations where they thought the device would not operate properly. However, with experience clips, we saw the users seeking novel usage situations and trying to push the possibilities of the device to its edge. The example clip "New interactions" shows the user exploring how the user interface works. We did not see this kind of use when we shadowed the users, as the users seemed to try to behave efficiently.

The field experiment suffered from several technical problems, which gave us the opportunity to explore the strategies and patterns of use that the users developed to overcome the technical difficulties. Sometimes these problems were turned into jokes. For example, the example clip "Dislocated cultural centre" shows a situation where the users have noted that the positioning service is not reliable or accurate enough, and they shoot a clip where they are in the local liquor store, but the positioning service tells them they are in the youth and cultural centre.

From frustration to "short films". The technique proved able to capture the users' emotional experiences. The clips revealed both spontaneous emotional responses to the system as well as small performances created by the users to express their emotions. An example of a small performance is shown in the clip "Scenario play", where the user seems to throw the device into the sea. Actually, he throws a rock, but the user continued the play even when he came back to return the device, explaining that unfortunately he does not have it anymore and showing the clip as evidence. When the observer was well-known to the user, it was natural for the user to explain their emotional responses and feelings towards the application, and they could then be simultaneously captured. Furthermore, we discovered that users expressed more lively responses and verbalised their thoughts more in the social situations with their friends compared to when they were alone or with a researcher with whom they were not familiar.

Some users seemed to want to avoid their failures being recorded. This was visible in some clips where they told the observer to stop recording. Some observers stopped filming, some continued. However, most users were quite happy to elaborate on their failures and negative experiences as well. The frustrations sometimes resulted in shooting something like the clip "Scenario play". When the users are frustrated enough, therefore, they are not satisfied with merely recording their true experiences with the applications, but they may begin to stage plays and shoot these on video.

New design ideas. The experience clips provided new design ideas, which seemed to emerge from the contextual influence as well as from the failures, or disappointments. For example, in the example clip "Visioning new features" a pair of our young student users were walking by a popular nightclub called "45", and when seeing it they had the idea that there could be a web camera filming the entrance of the nightclub and they could use the mobile city device for checking how long the queue was to support the decision whether to go the nightclub now or later. There were also clips created in response to disappointments that described how the users had hoped the application would operate.

Our findings were used for initiating changes that would solve problems revealed or improve the functionality and usefulness of the application. The study helped to identify the valuable directions towards which the design was to be developed. These included clips that showed the users' own ideas as well as those that displayed the apparently fluent and engaging interactions. The experience clips were rooted in the real-use context, which enabled a detailed study of what the users thought were valuable services while being on the move. Moreover, it helped to understand if the proposed design was able to provide the services in desirable and comprehensible form. The clips provided new ideas that resulted in added functionality and features in following design iterations.

In addition to the discovery of the potential of the designs, the study also provided us with important insights into the trouble with the application. For example, the design of the search functionality was proven be insensitive to the ways users wanted to conduct the search. ∎

Attitu

3

The manager is inspecting the plant

'You can have this one too. 'Björn writes in his book' ?!

Making sense and editing videos

AUGUSTO BOAL

"...the protagonist's body moves,
and this movement is in itself a writing.
This writing can and must be read."

3

Making sense
and editing videos

*The weather in each fjord is different. It's clear here at the moment but
in the neighbour fjord it may be snowing now.*
> – Freeride Skier in Lyngen, Norway, 2003

Is this information relevant to design? The quote is taken from a video recording of a skier unpacking his equipment at a roadside location in northern Norway. It was part of the freeride skiing study described in the previous
chapter. Without knowing the context (of both freeride skiing and the design
project) it is impossible to decide whether this bit of video is worth further
thought or can simply be discarded.

In another clip, the skier unzips his jacket to check that his avalanche
beeper is working. Is this a "problem" that reveals a design opportunity: a
beeper that does not require you to open your jacket? Or is this evidence of
a user value: the skier's concern for safety? Or a reminder that designs for
this target group need to comply with heavy winter jackets? Or is the clip
a trigger for the idea that skiers already have a power source, which could
potentially be utilised for other purposes? Interpretation depends on one's
interests as a designer.

When returning from video field studies the design team will face an
overwhelming amount of potentially relevant data. The user study may produce hours of videotape in addition to handwritten notes, photographs, sam-

ples of artefacts, users' documents, *etc.* Often the team members have visited different sites and thus have a different relationship to the material. Moreover, the personal differences stemming from professional backgrounds as well as individual preferences align the team members to focus on varying aspects of the content. What is considered most relevant to design and worth pursuing in further study depends heavily on these issues.

How can we share material and experiences, how should we focus, and how can we make sense of the material in view of the design task at hand? How can the "moulding" of video artefacts help propel the designers' and users' collaborative creativity and ensure that product ideas fit the users' reality? These questions will be addressed in this chapter.

The art of interpretation

Human behaviour – how people move, respond to events, and how they interact through talk – calls for designers' sensitive reading in the phase of interpretation. Video plainly repeats what it records. It is a tool that holds the fabric of life apparently intact for human perception.

> *Doing ethnography is like trying to read (in the sense of "construct a reading of") a manuscript – foreign, faded, full of eclipses, incoherencies, suspicious emendations, and tendentious commentaries, but written not in conventionalized graphs of sound but in transient examples of shaped behavior.* (Clifford Geertz, 1973)

There is a special kind of originality in video compared to symbolic materials such as texts and diagrams. Video preserves action in a sensitive and detailed fashion in relation to what originally happened. This allows subjecting the events to close scrutiny and enables designers to construct a deeper understanding of the timely interdependence and interaction between things. Geertz (1973) highlights the importance of paying attention to the timely organisation of an event by arguing that:

> *...the fact that this happens now, as opposed to then (whenever that may be) is crucial for providing some of the sense (in terms of context) for the event. Within the flow of action or interaction the notion of how actions relate to previous actions and preface future ones is essential to understanding.*

When interpretation is mediated by video recordings designers need to read the abundance of digitally reproduced stimuli, make sense of them, and describe them in new forms – like anthropologists who write ethnographies. In this process video recordings become meaningful assets to designers and help drive design discovery. Interpretation is a complex and multi-layered endeavour. Geertz (1973) stated:

> What the ethnographer is in fact faced with ... is a multiplicity of complex conceptual structures, many of them superimposed upon or knotted into one another, which are at once strange, irregular, and inexplicit, and which he must contrive somehow first to grasp and then to render.

The above quote highlights the difficulties in creating the so-called "thick descriptions" of studied cultures. The complexity of the idea is elaborated in Gilbert Ryle's (1968) simple story about two boys interacting with their eyes. One of the boys accidentally twitches. He does this with no attempt to signal anything specific to anyone. As a response, the other boy deliberately winks back. Both acts of contracting one's eyelid appear similar to the eye of the video camera, but their meaning is quite different to their owners and to the group of other people present.

According to Ryle (1968) a wink features at least five different levels of meaning: it is (1) deliberate communication, (2) targeted at a specific person, (3) with a particular message to convey, (4) according to a socially set code, (5) without the cognisance of others. However, a person winking is not doing five separate acts, but one. This single act is what a video record provides to the interpreters. The process of interpretation is a dynamic development of interpretations that are formed over one another. According to Ryle (1968) "... thick description is a many-layered sandwich, of which only the bottom slice is catered for by that the thinnest description." So, in Ryle's terms, video is actually a "thin description" – the bottom slice of the "sandwich of meaning".

Designers interpret user materials to drive designing. Thereafter, design interpretation needs to consider both the issues related to understanding the studied communities of practice as well as develop a sense of the different levels of meaning that products play in people's lives within the material ecology of products. In sum, design interpretation calls for the capacity to identify patterns that transcend individual observations of human interactions, the skill to build new ideas on these, and the ability to relate the whole to a design project's aims.

Analytic and empathic interpretation

Interpretation underlines the centrality of the idea of *meaning*. When designers attempt to understand how situations become meaningful to the people studied, they are working on the basis of the fundamental assumptions outlined by Herbert Blumer (1969, or 1986, p. 2):

▸ *Human beings act toward things on the basis of the meanings that the things have for them.*

▸ *The meaning of such things is derived from, or arises out of, the social interaction that one has with one's fellows.*

▸ *These meanings are handled in, and modified through, an interpretative process used by the person in dealing with the things he encounters.*

These meanings arise both out of the materiality of the situation and out of the biologically and culturally developed mental structures that guide how people perceive things. Thus, a mere analytic observation of details as objective facts would be too narrow an orientation for the study of the living contexts of use. People are sensual, emotional and experiential beings in addition to rational actors. Psychologist Jerome Bruner (1986) contends that, as humans, we have two modes of cognitive functioning, each of which have their own operating principles and own criteria for verification. He exemplifies the difference (1986, p. 11):

> *A good story and a well-formed argument are different natural kinds.*
> *Both can be used as means for convincing another. [...]*
>
> *One mode, the paradigmatic or logico-scientific one, attempts to fulfil the ideal of a formal, mathematical system of description and explanation. It employs categorization or conceptualization and the operations by which categories are established, instantiated, idealized and related on to the other to form a system. [...]*
>
> *The imaginative application of the narrative mode leads instead to good stories, gripping drama, believable (though not necessarily "true") historical accounts. It deals with human or human-like intention and action and the vicissitudes and consequences that mark their course.*

Analytical logic, as that employed in scientific endeavour, and verisimilitude, such as what engaging stories convey, constitute two radically different arenas for interpretation. According to this insight, relying only on a rational analysis of observable facts of human intercourse would be walking half-blind. This holds true both for ethnographic as well as for design interpretation. Jean Rouch (in Macdonald and Cousins, 1996, p. 266) promotes this in his remark on a film by Sergei Eisenstein:

> *The best film on Mexico is Eisenstein's Que Viva Mexico. Now, it happens that this film is completely false – it was all created, there wasn't one real scene in it; and the Mexicans themselves recognize it as the truest film on Mexico, simply because the fiction that Eisenstein reconstructed was closest to the Mexican image.*

Theatre director Augusto Boal has also observed that what is *true* to people may appear rather different from how things look in nature. He based his Image Theatre (Boal, 1992) on people's expression of the true character of, for example, their leaders. The *image* of how people in a particular culture see themselves may in some cases be more important than the facts in interaction. Reading the image and rendering it for others to read are crucial to the process becoming more conscious of the relevant issues. Jean Rouch (in Macdonald and Cousins, 1996, p. 266) continues:

> *I think that to make a film is to tell a story. An ethnographic book tells a story; bad ethnographic books, bad theses are accumulations of documents.*

How should designers then approach video recordings? The challenge is, on one hand, the conceptual and analytical study of patterns and relationships, and on the other hand the empathic reading and construction of images and stories of meaningful everyday life. The mixture of analytic reasoning and sensual experiencing in perceiving and conceptualising meanings makes interpretation an art in itself. Video has the capacity to serve up details for analytical scrutiny as well as to provide verisimilitude that fosters empathic engagement with people and situations. The malleability of video supports the development of insightful and provocative design artefacts; these in turn fuel the discovery of new perspectives on people's everyday existence.

Shared focus

Our senses and minds are developed to make meaningful observations of the world considering our life, action and intentions. Ryle (1968) observed that even a simple activity involves a great number of layered and culturally attuned skills to interpret the activity. This fact highlights the dramatic influence of an interpreter's personal knowledge and orientation to the process of interpretation, and the role that the interpreter's background plays in perception. This influence is also underlined by Blumer (1986, p. 36):

> *Whether we be laymen or scholars, we necessarily view any unfamiliar
> area of group life through images we already possess.*

In addition to focusing on different issues people perceive things differently. A cook has a sharp eye on how the person in the video handles the onions, and the aptitude to evaluate the skill of the cook based on the equipment she uses. An experienced interaction analyst identifies and is ready to express the subtleties of "participation structures" with an established vocabulary. The usability expert may point out the problems in handling the bowls. These examples stand to highlight some of the differences.

Attention also becomes affected by the interests of the current project. Like pregnant mothers who begin to notice surprisingly many other pregnant women around them, designers in a particular project become sensitised to the issues relevant to the intentions of the project. For example, in a "kitchen container design" project, attention would be drawn towards the interaction with various containers and situations around storing, moving and fetching ingredients. To the contrary, in the "mobile digital kitchen" project, the focus would turn to the information flows in the kitchen. The observations, and hence interpretations, that we make of the video recordings are inevitably coloured by our professional and cultural backgrounds, current intentions, as well as personal abilities and aptitudes.

This is where the collaborative process of interpretation provides its value. When different observations become the subject of discussion within a design team, these differences are brought to light. Shared interpretations help a design team open up new perspectives in looking at the material and find new opportunities for design. Interpretations are in this respect similar to "concepts", as explained by Blumer (1969, p. 160):

> *A concept always arises as an individual experience, to bridge a gap or
> insufficiency in perception. In becoming social property it permits others
> to gain the same point of view and employ the same orientation. As such
> it enables collective action – a function of the concept which, curiously
> enough, has received little attention.*

The re-orientation that new interpretations and concepts enable may not be
foreseen before the concept is discovered and shared. A linear process that
expects a progression from one stage to another in a sequential manner does
not account for the radical change in focus that a new concept may instigate.
An innovation process needs to shift from a linear sequence, where the ideas
are first sketched, then refined and implemented, into a parallel and cyclic
dialogue, where weight is put onto the formation of new insight.

According to user-centred design experts, design teams should feature
people with varying backgrounds and expertise to be truly innovative. For
example, Keinonen and Takala (2006) suggest six different roles that should
be represented in a design team: the user, the domain, design, communi-
cations, and feasibility specialists, and the team leader. Kelley (2001) is not
satisfied with six roles but proposes ten "personas", who focus on themes
of learning, organising and building. On the one hand, the challenges in
modern development projects are simply too big and too multi-faceted to
be handled by a single individual. On the other hand, differences in perception presume collaboration.

It is crucial, then, to establish a constructive dialogue among the various professionals, designers, engineers, managers and users, and between
interpretations and presentations. A firmly based understanding of the use
context helps ensure that designs will fit into users' reality. An analysis of
technology trends with engineers can facilitate focussing on ideas that can
actually be realised. The social and economic trends brought to the process
by managers and other partners help develop a sense of how well the product may compete with other possible solutions at their disposal. Dialogue
and co-construction of design visions is the pre-condition to effectively discovering the valuable issues in designing for people.

Pleasurable and effective co-interpretation

Video, as a highly communicable medium, provides diverse people with the
chance to contribute to interpretation. However, while video may enable everyone's full participation in collaborative exploration of detailed empirical
data, effective co-interpretation calls for additional support. Action needs to
be taken to orient people to observe the video materials with appropriate sensitivity and background. The participants in design sessions need to establish
a clear focus in order to make interpretations relevant to a design project.

As people may come from various backgrounds, and may not know each
other, attention needs to be placed on establishing a hospitable and safe

environment for expressing interpretations openly. This may require some initial warm-up activities in the beginning to help participants feel more comfortable with each other. At the very least, a design event should briefly go through who is who. People's backgrounds may help others to understand the kinds of interpretations someone makes. Moreover, it may help people with similar interests to get to know each other, when, for example, they explain in the workshop their motivations to participate.

When people feel that they are listened to, instead of evaluated, it is likely that collaboration will be fruitful. By encouraging the participants to build on each other's ideas and interpretations, the design events fuel an effective, collaborative construction of new interpretations. The feeling of being listened to may be crucial to enabling the participants to release their creativity in interpreting the video content. It may also be the most naive interpretations that help highlight new opportunities for development.

To help move beyond the initial impressions of the video content, thinking about how the video contents are related to each other needs to be encouraged. The high-level aim of interpretation is the discovery of new structures that explain and argue for new design opportunities. To ensure that the interpretations are also relevant to other situations than those in a single video clip, the team eventually needs to establish a broader perspective from which to look and reframe what they have already identified. As a result new insights and deeper questions may arise.

Interpretation may deeply affect later activities by guiding what is seen as important, in what direction ideas will be developed, and what activities will be supported by the designs. When collaboration is facilitated with proper focus on the participants as humane actors and on the goals of the project, co-interpretation has a fair chance to advance the project towards innovative products. Positive and memorable experiences have bearings on what is brought into important design decisions. In summary, the humane aspects matter both for the effectiveness of the process as well as for developing a sustainable and empowering atmosphere for the work.

● Method: Interaction Analysis Lab

*"What do
we see
here?"*

The Interaction Analysis Laboratory turns video interaction analysis into a collective practice. It was developed at the Institute for Research and Learning and at Xerox PARC as a practical way to encourage the use of ethnography in daily settings (Jordan and Henderson, 1995).

Although in its original form the Interaction Analysis Lab did not focus on design *per se*, it holds great potential for design teams that employ video to make sense of field studies in user settings. When introduced, it expanded prevailing video analysis practices on two core issues: it suggested a practical format for turning video analysis from an individual activity into a collaborative one, and it showed how researchers can find meaning in the video data grounded in the material itself rather than through applying preconceived schemas (like task analysis).

Interaction Analysis Lab can be organised as a permanent forum that meets in weekly sessions to jointly analyse video recordings. It is a forum where researchers from different projects can meet and help one another analyse material. Jordan and Henderson stress the importance of the group being multidisciplinary, as a diverse group will help "reveal and challenge idiosyncratic biases on the part of individual analysts" (Jordan and Henderson, 1995).

The Interaction Analysis Lab session runs for a couple of hours and involves viewing and discussing a video recording. The "tape owner" who brings his or her material to the session introduces the context of the recording, may suggest a particular focus for the analysis, and decides from where to start the tape. Once the tape is running, participants can say "stop" at any time to voice an observation or a hypothesis about what is happening in the recording. When the theme is exhausted, the tape moves on until a participant picks up on a new thread.

Jordan and Henderson stress that lengthy, speculative group discussions should be discouraged, as they tend to shift attention away from the actual data, from what can be seen and heard on the video tape. To make certain that the discussion stays on course, the "tape owner" may call upon participants to base their arguments directly on the material at all times. Or he may introduce the simple rule that the tape can never be paused for more than a few minutes at a time.

In practical terms one cannot hope to cover much more than 30 minutes of video recording in a two-hour session. Video interaction analysis in a mixed team is exciting, as it brings unexpected perspectives to the material, but it is also exhausting.

The Interaction Analysis session may use several tricks to guide sensitivity to particular aspects of the video. For example, turning off the sound while viewing encourages a strong focus on what is visible, rather than on what explanations one might seek in the audible dialogue. The handling of artefacts, body movements, and facial expressions gain significance. Another twist is to run the videotape in fast motion. This draws attention to the rhythm and periodicity of the practice being observed.

One way of reducing the tape owner's workload when analysing the session's outcome is to involve the full group in a post-it or sketching exercise. When asked to note down what they found most significant in the video recording, and subsequently arrange the post-its in an affinity diagram, the group contributes to building a representation of the new understanding of the material. Such a representation may in itself turn into a valuable design artefact.

Finding foci for the analysis

To assist interaction analysts in making sense of what they see on video, Jordan and Henderson suggest a list of foci for analysis that – without imposing a certain structure on the material – provides the untrained observer with a way of building understanding of human interactions. The list is generalised from the experience of analysing video in a range of projects at the Institute for Research on Learning and Xerox Parc (Jordan and Henderson, 1995).

The structure of events – Although human activity progresses continuously in time, people themselves will experience what they do in terms of beginnings and endings of events, and different segments of events.

The temporal organisation of activities – What is the rhythm and periodicity of the observed practice? Only if we understand the temporal structure can we observe when things break down – and possibly offer a design opportunity.

Turn-taking – It is an important concept from conversation analysis that people take turns speaking. Interaction is even more complex, as it includes ways in which people shift body postures, hand over artefacts, *etc.*

Participation structures – How do people group, who links with whom, who collaborates, and what are the formal and informal hierarchies?

Trouble and repair – When breakdowns or "trouble" in the regular activity occur, people take measures to "repair" the flow of activity. How do they do this?

The spatial organisation of activities – People occupy space in characteristic ways and the way they take possession can be very significant to, for instance, their role in a group.

The use of artefacts and documents – This focus is probably most central in analysing how users interact with technology. How, for instance, do people handle non-electronic artefacts compared to electronic ones?

This list may not be exhaustive, but it provides designers with a set of perspectives to breaking down a complex activity, and "handles" in the form of terms and concepts for talking about what one observes.

The strength of the Interaction Analysis Lab is that it provides a setting for in-depth discussion of video footage. Assembling a multidisciplinary group makes it more likely that multiple interpretations of the material will surface, making it easier to relate the footage to a design agenda. The format is, however, suited to shorter lengths of video recordings, as it is very difficult to sustain concentration for longer than two hours. ■

Interpretation as design

Interpretation functions as the glue that binds together realism and fiction – observations and visions. The fundamental paradox in design interpretation is that it needs to build both on what exists and what does not exist yet. This makes design interpretation challenging and exciting. Moreover, the fact that designers constantly work under heavy time pressures in industrial projects makes the challenge appear almost absurd.

For these reasons design ethnography and interpretation must radically simplify and cut down analytical rigour. Designers are forced to adopt a "creative" attitude during interpretation. Practically this means that the interpretation of certain video material becomes heavily influenced by who is interpreting and for which project the interpretations are made (see Figure 3.1).

Grounded or framed interpretation

Designers may choose from two basic approaches to interpret user data: *the grounded* and *the framed* approach. The grounded approach is an open mode rooted in the close study of contextual data, but does not impose any *a priori* structure on the data. The framed approach utilises a template or a model for the interpretation. The approaches differ in how they guide the exploration, interpretation and description of the material. To understand the impact of the differences on the quality of interpretations we need to take a step backwards and look at the fundamentals of interpretation.

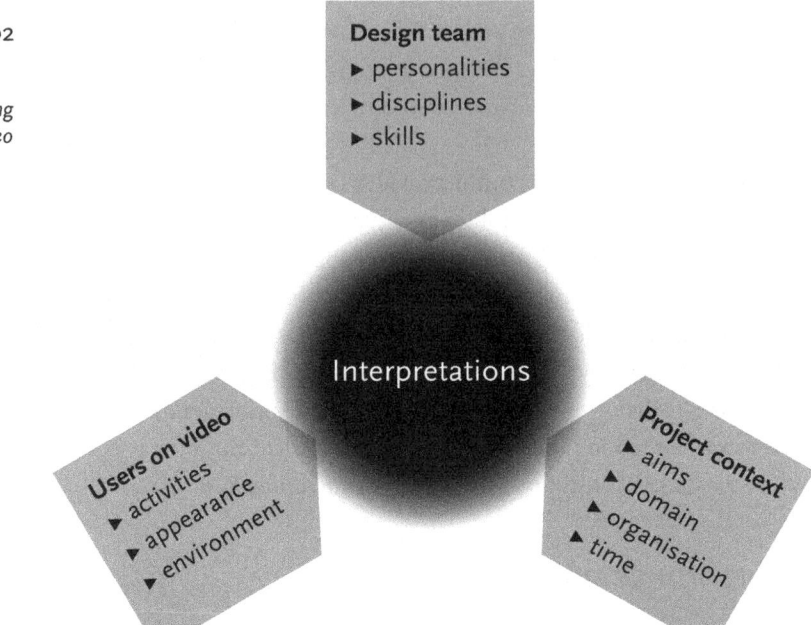

Figure 3.1
Interpretations are created in relation to the project context, use context, and the design team's characteristics

Interpretation forms a miniature model of a design process. It comprises the activities of exploring, relating and creating materials. Exploration is a means to form acquaintance with something that the observers are unfamiliar with, or which is completely unknown to them. Exploration helps to ensure that the interpretations and understanding of the design-relevant problems and opportunities "arise out of, and remain grounded in, the empirical life under study" (Blumer, 1986).

Exploration is flexible. It does not follow any specific protocol and is not fixed to any particular set of techniques. It may radically shift focus and move into new directions as these become encountered. During exploration, what is considered as being "relevant" may change completely. The explorative phase emphasises the ability to constantly challenge current views about the issues under study. Ryle (1968) brilliantly characterises the paradox of exploration in his example of an exploring traveller:

> *The paces that had taken him to the quagmire would have been a traveller's bad investment, but they were, on a modest scale, the explorer's good investment. He had learned from their fate, what he had not previously known that they would have been and will be a traveller's bad investment.*

Relating, on the contrary, forms a different orientation towards the data.

Blumer explains the procedure of *inspection* as a more focused endeavour
to discover a deeper understanding of a more constrained set of items under study:

> One goes to the empirical instances of the analytical element, views them
> in their different concrete settings, looks at them from different posi-
> tions, asks questions of them with regard to their generic character, goes
> back and re-examines them, compares them with one another, and this
> manner sifts out the nature of the analytical element that the empirical
> instances represent. (Blumer, 1986, p. 46)

The grounded approach borrows from anthropology and certain traditions in
sociology, especially that of symbolic interactionism. The basic assumption
is that the meaning and structure of the interpretations must emerge from
the data itself, as frames will always originate in a context (or culture) different from the one you are studying: "You must build on what is there, not
on what you have brought along", recommends the great Norwegian anthropologist Frederick Barth in a portrait film by Werner Sperschneider (2000).
Blumer (1986, p. 37) also emphasises that one of the biggest mistakes that
can be made in the study of people's social life is to let earlier concepts and
beliefs of one's own tradition serve as the substitute for firsthand acquaintance of that particular sphere of life. Making sound, sensitive and insightful interpretations is hard work and requires time and patience; and this is
often what is missing from a project. Time efficiency matters especially for
the sake of reduced costs in commercial design projects.

Framed analysis grows out of pre-conceived understandings, and helps
to make interpretation a more straightforward activity. While guiding observation, such frames state that, "this is interesting, observe this aspect of
the action or the environment". If such a model is available already upfront
at the stage of conducting the user studies, it helps designers by proposing
a clear framing that will produce appropriate data for the models. Some established models also delineate how to describe the findings and thus save
designers valuable time. Models such as in Contextual Design (Beyer and
Holtzblatt, 1998) bring forward examples of how to draw analytical and abstracted pictures of artefacts, roles and information flows, physical spaces,
hierarchical structures of activities, and the cultural forces active in an organisation. The models are based on years of user-centred development of

information systems and are thus well tested and have been proven useful in a great variety of IT projects.

However, domains vary drastically in what is important for design, and design projects vary greatly in their aims. Workflow diagrams and task hierarchies are well suited to the analysis of the types of work where information handling is the primary purpose. They help designers break down the overall process into sub-processes and tasks, which may eventually be supported or taken over by IT systems. However, for a design team exploring other domains and non-work settings, the models may not prove as helpful. For example, in the "Freeride skiers" case the challenge was to explore opportunities for creating new sport equipment for freeride skiers. The ideas may include information devices – but information handling is never more than secondary to the actual experience in freeride skiing.

Interpretation is hard work, but it is also handiwork. Interpretation such as that done by interaction analysis (Jordan and Henderson, 1995) is best learned by doing. Although Jordan and Henderson provide designers with definite foci, the analysis relies heavily on the expertise of the participants. In a similar manner as how maintenance technicians eventually grow sensitive to listening to how the machines "talk" when making the diagnosis for repairing them, designers may develop richer knowledge in identifying various participation structures and increased sensitivity to the subtleties of the temporal organisation of activities. The essence will vary depending on the project, and it belongs to the workmanship of the interpreter to choose the appropriate method – whether it is grounded or framed. Ultimately the choice between a framed and a grounded approach is between the qualities that Dewey (1910) discussed:

> *Projection and reflection, going directly ahead and turning back in scrutiny, should alternate. Unconsciousness gives spontaneity and freshness; consciousness, conviction and control.*

The apparent fluency that framed models provide reduces sensitivity to the differences between domains and projects. When a design project aspires to create radically new ideas it is obvious that fixed models tend to promote too rigid a perception of phenomena. Sometimes results are needed quickly, and such models help to achieve convincing results in a rather short time. Time is precious and how it becomes invested in the phase of interpretation may have tremendous impact on the phases to come. Sensitive interpretation resides at the heart of good design, and with the use of video designers may develop a greater sensitivity to grounding design on true images of reality.

Figure 3.2
Video card game players constructing themes with video cards

● Method: Video Card Game

The video card game lets a design team cover a bulk of video material in a few hours by segmenting it into smaller chunks. It was developed in the Danfoss company to enhance collaboration between user-centred design consultants and engineering development teams and to encourage the development team to take ownership of user problems with their products or prototypes (Buur and Søndergaard, 2000).

"How are these videos connected?"

In the early phases of the design project (field study and interview video), the team focuses mostly on making sense of the material and forming early ideas. The video card game typically results in – often surprising – perspectives on the material: issues worth exploring further and design opportunities that may be investigated. In the later project phases, when prototypes exist (workshop and usability evaluation video), the focus will be on identifying problems, prioritising them and finding solutions. The game encourages a focused understanding of which problems need attention.

Like Interaction Analysis Lab the video card game banks on the involvement of people from different disciplines to make sense of the video.

Figure 3.3
"Reading the
video cards"

It differs in that it works with large amounts of video (several hours) cut into short video clips. It also combines individual viewing with shared sense-making. It is especially suitable for comparisons of material recorded across several sites. The method works effectively with a broad range of video material, from user observations in the early stages of a design project to usability evaluations in the later stages. It is based on a bottom-up approach to interpreting field observations.

The video card game took inspiration from the "Happy Families" children's game. In Happy Families the players each try to collect families of four cards (for example "cats" or "dogs"). They do this by asking each other in turn for cards ("I'd like a dog from you"). In video card game, a set of picture cards represent the video clips and allow participants to handle them as in a card game: spread them out, group them, form series and patterns, and exchange them with other participants. The cards focus discussion on the video material and how the participants interpret it. By turning video analysis into a delightful and fun game that – even so – provides convincing insight, the method captures the attention of fast-paced design teams.

A typical one-day video card game session uses 30 to 80 short video clips with approximately 10 game participants. The participants are seated around a large table with video equipment. The session starts with an introduction to the video recordings (where they were taken and by whom) and the goal of the game. The procedure then follows the steps outlined below.

Step 1: Dealing the cards (30 min)
The cards are dealt randomly between players and the rules of making observations are explained. Random selection helps the players focus on the contents of each individual clip. Having different materials available also helps to trigger ideas onto a broader track.

Step 2: Reading the cards (1 hour)
The players then split up to watch their video clips. It is enough to watch the clips in this phase only once or twice and make quick notes that describe observations made about the clips. By annotating each card in their own handwriting the players come to "own" the card, which is important in the later

stages. If players work in pairs the fact that they share one card forces them to discuss what they have seen and formulate observations together.

Step 3: Arranging your hand (30 min)
When participants return to the game table they are asked to group their cards openly in front of them on the table. This encourages the players to start forming ideas about what might be important to them in the clips. Each player around the table briefly presents his/her structure. There are no restrictions on how players group their cards as long as it makes sense in terms of the design activity (*e.g.* user activities, design problems).

Step 4: Collecting card families (1 hour)
Each player (or a pair of players) is then asked to choose their favourite family of cards. One after another the players describe the theme they have chosen as precisely as they can. This invites the other players to contribute with cards that seem to fit into the same theme.

Before moving from one theme to the next, the facilitator mounts the cards belonging to the theme on a separate poster. Collecting the card families continues until all cards have found a place in the structure. The grouping of cards encourages discussion on finding the exact wording of the theme heading; it needs to be precise enough to define which cards belong and which do not. Sometimes cards can make sense in more than one group. In this case a blank card serves as a duplicate with an index reference.

By selecting their favourite themes, the players also take responsibility for a theme including the labelled poster with cards. This helps in the later phase to jot down findings, when collaborative observations are made about the clips belonging to the theme.

Step 5: Discussing the card families (3–4 hours)
To gain an overview of the themes, the theme posters are pinned to a wall board or projected with a data projector. This provides the opportunity to reflect on the immediate outcome of the game. The participants are then asked to arrange and prioritise the themes: Which one do we need to discuss first? Which themes seem most important to the design project? The players discuss the families one after another. Each "theme owner" is encouraged to lead the discussion and add notes to the poster. Since none of the players have seen all the clips, it is important to return to the video at this point. Typically each player will show and explain "their" clips to the oth-

ers, and argue how these clips are able to increase understanding about a theme. Sometimes the players will want to see the video repeatedly throughout such a discussion.

Mock-ups, prototypes, and artefacts collected in the field have proven to be good facilitators of the discussion when they are readily available on the table to point at and think about. They help guide the discussion towards design ideas and hence help to construct a relevant focus for designing. The video card game can lead beyond mere interpretations of the material to team decisions on how to move forward and what to do next. The video cards also serve as "tangible arguments" that can increase participants' confidence when they present and argue for their new ideas.

At the end of the video card game, the immediate results – the posters with video card themes and notes – are copied and circulated amongst the participants. Often this simple documentation is sufficient for team members to be able to prioritise activities and divide tasks among them for the next design move: Who should further investigate what, or which design problems need attention.

Preparing video material and cards

The video card game works best with video material that contains visual activities, *i.e.* communicates on a non-verbal level (field observations and usability evaluation videos). The idea of the game is largely to turn the visual into verbal, and to make it subject to a design discussion. Video recordings that are dominantly verbal, such as interview and discussion recordings, do not necessarily need a video-based approach. These materials can be interpreted with verbal methods, such as affinity diagramming. Hence, if designers plan to utilise the method, they need to keep in mind that observations should not turn into interviews during the field studies. Making successful field observations is discussed in the previous chapter.

Figure 3.4
A video card

In preparing the video clips and the cards the ones who made the recordings go through their material and select clips that show the most significant actions. The clips are typically thirty seconds to two minutes long and preferably contain one closed event rather than many. There is no particular principle for selecting clips. Designers will go by their professional interests, *i.e.* they can pick what they find puzzling, surprising, characteristic and otherwise relevant to the

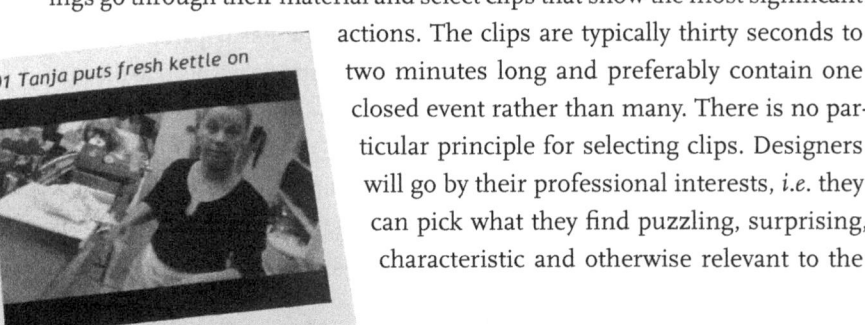

project in focus. In this phase they will not be expected to explain their choice of segments. The video segments will inevitably trigger observations beyond what the researchers can imagine; hence the selection of video will not steer the discussion in a very specific direction. Rather, the videos delimit the field of exploration: one cannot expect participants to talk about what they cannot see.

The number of clips will vary depending on the material and on how many participants there are in the game. The card game usually works best with 30 to 100 sequences, and each participant can handle 10 to 20 cards in a reasonable time for making observations. The video clips should be available in digital form so that they can be watched in an arbitrary order; any computer editing software will do. To strengthen the link between the clip and its card, they need to be named consistently.

The naming of cards and clips is significant as it influences the flow. The name of the person(s) depicted encourages empathy (*i.e.* it makes a difference to talk about "Lars" rather than "this person"), and the activity description should be neutral and brief – to avoid suggesting a particular interpretation. Numbering the clips makes it faster to refer to a particular clip in the heat of discussion. When more than one person prepares the clips and cards in parallel this means deciding on a numbering system upfront. Preparing the video clips and cards takes time – do not procrastinate!

Setting up the game table

The way the room is arranged for the video card game and how the equipment is placed has a remarkable influence on the dynamics and outcome of the design discussion. How do we position the table, chairs, boards, and screens? In the course of our work with video card game sessions we have experimented with several layouts, but we tend to return to a familiar meeting room or tight desktop-type setup.

A critical factor to a successful session is to find a layout where participants feel comfortable and can work on equal terms. Another is to make sure that participants can easily see and reach the cards. We have learned that the players will not employ video during the discussion if the spatial barrier to grab the card and play it is too big, or if they have to stand up in front of the group whenever they want to make a point. The players need to be seated within easy reach of both the cards and the monitor.

In addition to organising the space, the way participants are invited into the game as they enter the session affects how the game unfolds. To make

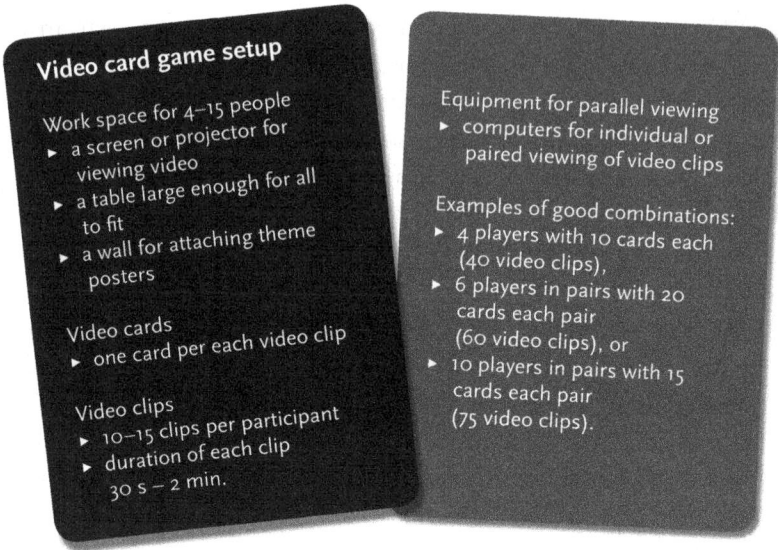

Video card game setup

Work space for 4–15 people
▸ a screen or projector for viewing video
▸ a table large enough for all to fit
▸ a wall for attaching theme posters

Video cards
▸ one card per each video clip

Video clips
▸ 10–15 clips per participant
▸ duration of each clip 30 s – 2 min.

Equipment for parallel viewing
▸ computers for individual or paired viewing of video clips

Examples of good combinations:
▸ 4 players with 10 cards each (40 video clips),
▸ 6 players in pairs with 20 cards each pair (60 video clips), or
▸ 10 players in pairs with 15 cards each pair (75 video clips).

the video card game work as *social glue*, it is important to establish a playful, yet goal-oriented, atmosphere from the start. We give the researchers time to talk about the people they have met and how the videos were recorded. As participants only get to see snippets of the full video material, it is important to provide some broader context.

With novice participants we prefer to start with a small interaction analysis exercise with an example card to sharpen the attention on visual content, to demonstrate how different people observe differently (and that this is beneficial), and to point out the difference between observation and interpretation. Observations are things we can actually see in the video frame: they do not need inference about what people think, or about what happened before or after. For example, an observation from a video clip from the kitchen project could be "The woman hands the girl a plate in the kitchen". It is something that no one can doubt when seeing it. An interpretation of the same clip could be "The daughter needs her mother's help in setting the table" – but we cannot see that she will be laying the table, or that she indeed needs help. Bold interpretations are left to the second round of the game. ∎

Themes that trigger design
Themes such as what the video card game constructs help to chunk material into more easily handled pieces. Themes usually become expressed with a

Figure 3.5
Mads and
Joanne dis-
cuss what the
person on
the video is
doing

▶ Case story: Video sensemaking
Jason Moore, University of Southern Denmark; now: Xinsight

Mads (an engineer) and Joanne (a nurse) have just viewed a video clip of
a diabetic injecting insulin in a café, as preparation for a video card game.
Faced with the challenge of describing their observations on a video card,
the following conversation unfolds.

Mads to Joanne	*So, now we should describe him?*
Mads	OK. *[Mads starts to write "Syringe" on the video card.]*
Joanne	*Even now I consider him a little alternative. I don't know why.* *[Joanne looks over at what Mads wrote.]*
Joanne	*"Syringe". [Pause.]*
Mads	*He's eating breakfast someplace.*
Joanne	*Yes.*
Mads	*It's a little... Why isn't he at home?*
Joanne	*That's right. Yes, he's such a... what can we call it: Eating breakfast out?*
Mads	*Breakfast. [Mads and Joanne laugh.]*
Joanne	*Café guest. [Pause, Mads starts writing.]*
Joanne	*Can we not just write that he... eats breakfast out?* *[Mads writes "Café-breakfast" on video card.]*

Joanne	*I'm thinking also about publicity, he really doesn't seem particularly shy. It seems to me that he is injecting himself in public there. I don't know, but it seems like it.*
Mads	*What is it called, public...?*
Joanne	*Public diabetic. You can explain it if they ask, right?*
	[Joanne laughs and Mads nods and smiles as he writes "'public' diabetic" on the video card.]

What is striking in this dialogue is how real-life video triggers numerous points of focus. Joanne initially focuses on the qualities of Brian (the person in the video), identifying him as "alternative", while Mads writes "syringe" indicating that he has focused on the fact that Brian uses a traditional syringe to inject his insulin (Mads designs insulin injection pens). He then focuses on the fact that Brian is eating breakfast, and wonders why Brian isn't eating breakfast at home. Joanne takes the focus of Brian not being at home and rephrases it to wonder why Brian is eating out in public. This leads her to focus on Brian's personality, and to comment that he is not shy, as he seems about to inject his insulin in public. She then comes up with the elegant phrase "public diabetic".

Mads and Joanne are members of a design team at Novo Nordisk, a Danish pharmaceutical company whose core business is developing products for the treatment of diabetes. Novo Nordisk approached us with an interest in a more user-centred approach to product design, and so we proposed ethnographic field studies to provide insight into the daily lives of people with diabetes. As we were entering people's private lives, we decided to work through video recordings rather than attempt to bring people in direct collaboration with the design team. The video card game was organised for the team to learn about their "users" by collaboratively analysing the video. The participants were mainly mechanical engineers who work with designing needles and injection devices, although there were also participants from marketing and clinical research. The goal initially was to get to know the people in the video, and later to identify design opportunities and envision new products.

It is obvious from the dialogue that different people see different things in the same video. This is the powerful quality of video: even short clips allow viewers to find multiple focus points. After the video card game Mads commented that he liked working with Joanne since she was a "personal" observer, where he was an "inventive" one, which together helped them to see "twice as much". The drawback is also evident at the start of this conversation, as Mads and Joanne have difficulty finding something that they both can agree is interesting. It is only when Mads wonders aloud why Brian is not eating breakfast at home that they share a topic. Note that it was not Mads' initial focus that triggered Joanne's reaction, but his reflecting on that focus. Wondering why something is happening is different from merely identifying that an event happened. By wondering, a point of focus is identified and selected as being important enough to be investigated further.

Focusing on certain details in a video clip is a natural and intuitive act, but simply pointing out what is interesting does not advance the design discussion. Points of focus must be explained as to why they are important to become topics of conversation. The task of writing on the video card is particularly useful in that it encourages participants to stop and collectively reflect on what is important in the video. This type of reflection is hard work, so it is important to structure workshops such that participants are encouraged to do this conceptual heavy-lifting. The video card exercise creates time for this type of thinking to occur.

In some circumstances, reflection on particular points of focus in the video leads to a more general reframing, or a new understanding of the design problem. In the following transcript, two mechanical engineers Peter and Claus are viewing a video clip of Cynthia as she prepares to make an injection in her kitchen. Michael and Hans are sitting at a computer on the other side of the table viewing the same video clip.

Cynthia [on video]	*I need to get... Sorry, I need to get another pen tip. [Walking to living room] Not very well... [looking in her purse] I have all these bags and bits and pieces of stuff... that I carry around...*
Jason [on video]	*This is where you keep your pen tips?*
Cynthia [on video]	*Yeah, normally actually they are in... [searching purse] I keep... [walking back to kitchen] I have a little place in my glucose kit... I keep in here extra cartridges of each, the NPH and the Novo Rapid, and at least two pen tips. [Video clip ends.]*
Peter	*Pen tips. [To Jason] That's the needles? [Jason nods.]*
Peter	*Pen tips. Never heard of it.*
Claus	*It's much nicer.*
Jason	*No? You don't call them that ever?*
Peter [to Claus]	*Yeah, it is.*
Claus	*We're so needle fixated.*

Peter	Needle sounds so drastic. [Turns his attention back to the computer.] Pen tips.
	[Overhearing this, Michael from the other side of the table breaks from a discussion with Hans to join the conversation.]
Michael	Why... Jason, why does she call it "pen tips" instead of needles?
Jason	I don't know. That's just the term she uses. And so then I just called them that also.
Hans	It's a good term.
Michael	It is.
Jason	I don't know if she invented it...
Hans	It's a very good term actually.
Michael	Especially if you don't like the whole concept of needles and injection, then it might make it more....
Hans	You don't have to say it at least. Needles.
Jason	You haven't heard that term before?
Michael	Pen tip? No.

The game participants all focus on the new term "pen tips". Partially, it is because they have never heard the term before, but the novelty of the term does not fully explain their interest. They note that it is a "good term", "much nicer" and less "drastic" than referring to "needles". It leads them away from their current "needle fixated" viewpoint to see that some people prefer to not even talk about "needles". This type of insight goes beyond identifying what is interesting or relevant in the video, as the participants are actually developing a new understanding of their design space. The term "pen tips" is a reminder that the people who use the needles have a different perspective from the designers, and it challenges the participants to reframe their understanding to include this new way of looking at the product.

Not all reframing is so immediate, or visible. An insight may remain a private thought, may only be revealed in a private discussion between two participants, or may not even be formulated at all. For this reason, the presentations that are part of the video card game are crucial to capturing the key results of the workshop, and encouraging participants to make their insights explicit. In the same way that the video cards encourage participants to reflect on their points of focus in a video clip, the presentations help to prioritise the key reflections and problem-framing ideas that span all the video clips. ■

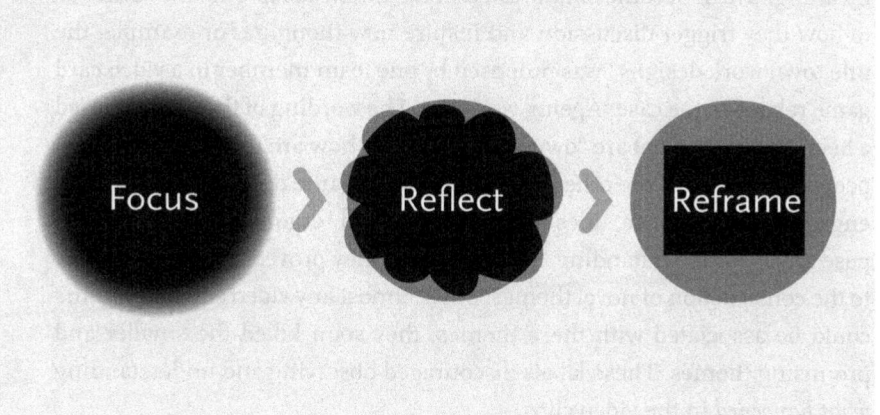

Figure 3.6
What hap-
pens when
designers
make sense
of video.

title and some concretising examples of the content. The way the title is articulated together with the content provided as its support constitute a powerful tool for coordinating the actions of a design team. It is fundamentally
similar to how Blumer explains how "concepts" function in social interaction.
The orientation that a good theme provides may open up new design opportunities. However, a theme may lead astray as well. Themes form crucial elements in a design team's efforts to establish a shared vocabulary that allows
them to explain the use context as well as to argue for new opportunities.

The philosopher Wittgenstein (1976) expressed the role of words in human interaction in his famous phrasing: "words are also deeds". In the collaborative construction of themes on video material this quality of expressed
words is central. The way that the themes are verbally constructed delineates
which pieces of the video content become associated with each other. When
the design team presents a possible label for a theme, it directly constitutes
a rule. For example, when the label "painting traces" was discovered on the
video card in the "Freeride skiers" case, it made the team forge a new pile of
video clips where the skiers carved curves onto the snowy surface of moun-
tains. It proposed a strong orientation to look at the video material to identify
how "painting traces" could be understood in the people's visible actions.
Hence, themes are not only defined and characterised by the names they are
given, but also *formed* as a consequence of giving a name to them.

Video clips do not carry a dedicated meaning, which makes their interpretation both challenging and rewarding. There are no correct or incorrect
answers. Through interpretation video clips are assigned with a negotiated

meaning, which becomes manifested in the theme label. The titles differ as to how they trigger discussion and inspire new thought. For example, the title "own work designs" was proposed by one team member in a video card game related to the case "Ageing workers". The wording of the title triggered a hectic debate – what are "own work designs", how are they visible in what people do? Some of the other titles in the same game did not prove to be as engaging. For instance, titles such as "tools" and "ergonomics" did not propose any new understanding, and moreover, they proved to be catastrophic to the construction of novel themes. Since almost any video clip in the game could be associated with these themes, they soon killed the smaller and promising themes. These labels discouraged observing and understanding *what happened* in the video clips.

In the fast-paced video interpretation session, the thematic groups may become negotiated largely based on the initial impressions and groupings of the participants. Due to the speed only some of the most interesting video clips are collaboratively viewed and discussed in detail in video card games. This increases the importance of early identification of themes that have the capacity of driving design further along a fruitful track.

A good theme implies sensitive observations of the video content. A good title makes sense to the designers and provokes discovery. Themes foster ideation, fuel design discussion, and bridge ideas to the field data. As provokers of associations titles may open up new opportunities for design by bridging domains. For example, the everyday activity of reading a recipe can be provocatively labelled as "navigating in food". This sudden change in perceiving the activity as something else helps to bridge ideas from another domain to discussing how cooking could be served by intelligent designs for navigation.

The following list characterises good design themes based on video material. A good theme title

- ▶ *describes the action* on video;
- ▶ exposes a *relevant insight* (such that once known, the design team may not proceed without it);
- ▶ bears *new knowledge* for the design team;
- ▶ *inspires* the designers;
- ▶ *sets a new perspective* on looking at matters;
- ▶ arouses *new associations*;
- ▶ *manifests a clear rule* for choosing content from the footage.

A theme may not meet all these criteria, and as ethnographic accounts, themes are never complete and may be constantly re-interpreted anew into themes of new kinds.

Designing video artefacts

As soon as the design team has studied the user material, discovered and settled on meanings, the team is faced with the challenge of conveying their understanding. This may be articulated as diagrams, drawings, sketches, videos, *etc.* The format of the articulation matters for how the understanding can be shared, used and internalised. Moreover, the presentation format influences which issues become effective as drivers for design.

Video presentations are able to invoke concrete images and sensations of the real actions, environments, people, sounds, and feelings that are much like the real situation, the source of the data. Video is often the most accurate retrospective account of real action that a design team possesses. The moulding of video material from field studies enables designers to craft effective presentations that embed the living everyday and, at the same time, convey a deep conceptual understanding. In a sense we may regard a video presentation as a "theory" similar to an ethnography (or ethnology):

> *Good ethnology is a theory and a brilliant exposition of this theory*
> *– and that's what a film is. That is, you have something to say.* (Jean Rouch, in Cousins, 1996, p. 266)

As catalysers of designers' aspirations to change existing situations into preferred ones the video presentations have a fundamental role in the process. To strengthen this understanding we will term them *video artefacts*. Video artefacts *link* field data to design ideas, *inform* about what is relevant, *generalise* findings by combining data, help to *empathise* with people, and *focus* design by directing the interest. They also help evaluate designs in the later phases of the project. They may influence the quality and relevance of the subsequent work, where effective designing needs particular attention. Hence, the video artefacts that result from interpretation sessions, such as the themes in video card games or new video presentations of the understanding, are in themselves purposeful *designs*.

The following list characterises the desired qualities for video artefacts:

Bridges the gap between "them" and "us". By expressing about how people cope with their challenges, how they look, and what they think, the video artefact conveys empathy.

Provokes designers to rethink their taken-for-granted views of "problems" and "solutions". By depicting how people act, what drives them to act, and what their basic dilemmas are, video artefacts provide designers with a broader understanding of their challenge rather than offering simple accounts of problems to be solved through design.

Provides supporting arguments for designers. By offering clear, crisp terms that designers can adopt, and by exposing causalities that are graspable and easily communicable, the video artefacts form a resource for action, for designing and justifying designs.

Allows evolvement through negotiations in the design process. Theories are never "complete" or "finished". The video artefacts should not only communicate findings, but should serve as a frame for discussion among those who have studied the field.

Is transparent in terms of who is interpreting and for what motive. The video artefact introduces clearly who has studied the field and interpreted the data, and with what intentions. It discloses who is paying and what the overall motive is.

The malleability and presentation power of video provides an effective means for designers to cast their interpretations into the video artefacts. Figure 3.7

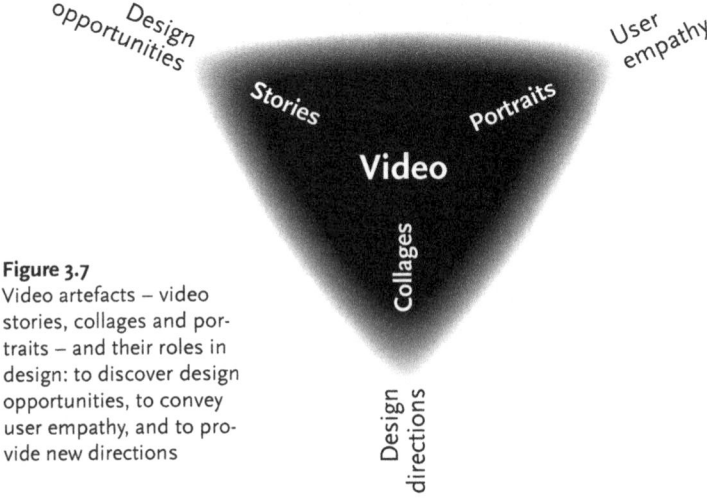

Figure 3.7
Video artefacts – video stories, collages and portraits – and their roles in design: to discover design opportunities, to convey user empathy, and to provide new directions

outlines how different kinds of video artefacts can be edited from the user study video footage to facilitate design discovery into particular areas.

A *video story* shows how things happen. It is an edited example that displays how a certain event unfolds in real life. It may be a combination of multi-camera observation, an illustration of a certain flow of actions within a larger setting, or it may be a detailed show about a particular user task.

A *video portrait* conveys empathy. It is a brief illustration of the design team's conception of a certain person. By combining the voice, image and activities of somebody, it helps to understand the way of life, attitudes and values, and the grounds that the person uses in making decisions.

A *video collage* provokes thought and facilitates the discovery of new areas, or directions, to move into. These areas may be concrete, such as certain activities or environments, or conceptual, such as the ideas of "group awareness" or "painting traces".

The making of video artefacts will be explained in the following method descriptions. The accompanying case stories help to convey how the video artefacts tend to merge into some form due to the particular characteristics of the material that designers are able to create in real projects. Often such videos are a mixture of these types (like most of the examples on the attached DVD). Figure 3.7 above is therefore most helpful if understood as an attempt to clarify possible aims of video-mediated interpretation.

It is a rather different conception originating from Keith Johnstone that the re-using of elements makes a "story", which is presented in Chapter 4, or the idea that a story needs to have an Aristotelian three-scene structure with a beginning, middle, and
† end.

● Method: Video Stories

Video stories describe how things happen.[†] The video footage is usually too heavy to be utilised in design sessions as such. It may contain activities that span across several user sites or contain material from several video cameras. Moreover, activities may unfold in parallel, and intermingle, which may make the "reading" of the material unnecessarily difficult for the audience. A shorter video story may depict some of the most interesting flows of activities or social interactions that have been captured on video, or draw attention to the skilful interludes exhibited by a competent user.

"Is it that complicated to change a lightbulb?"

A video story may help verify the understanding that a design team has built from a field study visit, or help ground ideation on the concrete situations during a journey. An incisive presentation of users' interactions may also underline the need to reconfigure the material environment. Sometimes it is the editing of the story that already creates a deeper understanding of the activities displayed and provides a design team with a new challenge.

Some tips for editing a video story

▸ Maintain continuity so that the flow of activities is easy to follow. If this proves too difficult (sometimes the material poses some challenges in this respect), use other ways, such as texts, fade-outs, or live explanation in the session, to ensure that the audience understands what is happening in the picture, how the activities relate to each other, and when situations shift.

▸ Keep field notes with exact time codes to later help fast location of the best clips.

▸ Use the "rec-pause" method with two interconnected video devices when in a hurry.

Video stories may illustrate how multi-faceted and complex even simple real-life tasks actually may be. For example, a sequence on a schoolhouse caretaker changing a lightbulb may conveniently illustrate how complex a procedure the "simple" task in real life actually is. The person needs to get the ladder from the cellar, the lightbulb from another place, then carry the ladder and the lightbulb without breaking the bulb, to the place where the burnt-out lamp is. Only now is the bulb "changed" through a small set of subtle acts. This is followed by a number of activities relating to moving the ladder back and disposing of the broken fluorescent bulb, according to appropriate hazardous waste procedures. The video sequence on this activity also describes this detailed contextual information, which would be lost if the task were transformed into a diagrammatic format.

Sometimes feedback from the users is invaluable in helping designers to find a proper focus for what should be improved. The "Operator feedback" case explains how a video story was used to help the designers grasp what was actually going on in the wastewater plant where they had observed the operators' work. The feedback enabled them to understand what they had missed – despite it being captured on video. In the editing phase the designers had omitted the skilled problem solving events, which would have been crucial to help them grasp the most burning issues to be improved.

The "Kitchen impressions" case explains the use of a video story (or a mixture of a collage and a story) to support the designers' orientation towards the activities in real kitchens. By providing concrete passages of real home

▶ Case story: Operator feedback

Jacob Buur, Danfoss User Centred Design

Video
example
**Do we
understand
the practice?**
2'00"

We are in the lunchroom of the local wastewater plant. The full crew is there: process operators, plant electricians, maintenance technicians, the lab technician, and the secretary, eight in all. We have brought morning rolls, and the agreement is that we have one hour to get feedback on the field video material that we recorded a few weeks earlier.

Kirsten, my colleague, introduces the video collages that we have edited for this meeting and controls the video player. I am on the other side of the room, using a small video camera on the table to document the discussion. For the meeting we have selected around 15 short video clips that we find it difficult to understand. Some of them we play one by one, while others are coarsely assembled in an edited video sequence. Kirsten plays each video clip, stops, and asks questions about what happens here. At first, the conversation develops somewhat slowly. The operators are a bit embarrassed about seeing themselves on screen and commenting about what they are doing. But once they realise that we are actually interested in finding out why they did what they did, they become more confident in explaining.

The Water Vision Project. The operator feedback session was part of the field study in a vision project on new technology for the water business segment, organised by the corporate User-Centred Design group of Danfoss. The goal of the project was to study the water treatment field from a user's perspective and suggest a vision for Danfoss products and user interfaces. The feedback session was the third meeting with the operators at the plant. The first time was a general introduction and walkthrough of what the plant was like, and the second time was the full-day field study with multiple camera teams each following their operator, as described in the case "Plant operators".

Did we get it right? One sequence in particular causes discussion among the operators. It is a pump repair situation that seemed to unfold with several people involved in observing, telephoning, discussing, and modifying. After the field study we were not even sure if this was something special, or just a routine repair. For this meeting we have prepared an edited version of what we thought happened in compressed form. The reaction we get from the head operator is:

– But you did not include the problem solving!

In his mind we have left out the most important – or challenging – activities in the sequence: the ones where they tried to understand the problem and came up with solutions.

Probably because we did not grasp the significance of the situation at first, we had focused on the manual actions and accidentally left out the part he is most proud of – the way they solved the problem quickly and efficiently with a good deal of ingenuity.

As most of the people around the table were involved when the "Holm Breakdown" (as we later labelled it) occurred, this sequence triggers a lot of dialogue about what each of the employees was doing at that moment, how they contributed, and what is missing in the edited video.

Building on the operator explanations we were able to edit a better account of what actually happened, and the sequence became quite influential in the subsequent design process. ∎

cooks it enabled designers to ground the discussion on design opportunities
for new technologies in real situations in kitchens.

Video stories may also help to convey an overview of the design project itself. Such stories have proved to be useful in design projects that have involved multiple stakeholders and that continued over long periods.[†] These video stories helped bring the atmosphere from previous events into the next ones, fostering a sense of involvement in a project with a great number of exciting events and people. This is a wonderful example of how video may function as social glue – inviting people to design together. ■

† These were utilised in the Luotain project cases Mobile Clinical Collaboration and Mobile Lurking and Kiteboarding in 2005.

● Method: Video Portraits

A video portrait is a video presentation that explains who someone is. It helps designers empathise with the users. Empathy is the ability of designers to put themselves into the users' shoes. It facilitates framing the design challenge so as to promote what users think is valuable. When portraits are authored with materials from user observations and interviews, the finished portrait often conveys a strong sense of the real people the designers encountered in the field.

"This is she."

User portraits have been utilised in design since Henry Dreyfuss developed concrete – but imagined – characters with the names "Joe" and "Josefine" to draw attention to the users (Dreyfuss, 1967). Since then, numerous variations of user portrayal in the design process have been developed, the most famous of which may be Alan Cooper's goal-driven personas (Cooper, 1999). These descriptions helped to discuss what the users needed and desired, and how the ideas should be adapted to better suit them. In this way they helped bring users closer and more effectively to designing.

Personas, however, were designs like products, and too often it happened that they did not have real relevance in the field. For example, Dreyfuss utilised drawn characters, and followed essentially a measurement-based approach that promoted ergonomics (or human factors) and anthropometrics, which neglected the messy interactions of daily practices. To the contrary, video portraits are usually created with design ethnographic video accounts, which ensure their relevance to the use context.

Editing a portrait may require that a design team is prepared already at the phase of capturing to make portraits. If the video is shot without the intention to create a portrait, it may later be quite difficult to mould the materials into a convincing and inspiring presentation capable of conveying a person's character.

Video
example
**Kitchen
impressions**
2'58"

▶ Case story: Kitchen impressions

Case author: Mette Mark Larsen, University of Southern Denmark

As in most Danish households, Amanda and Peter always have dinner in the evening. Being a young couple, they cook together – today they are making chicken fillet vegetable mix and a carrot salad. Peter stirs the meat and adds spices, while Amanda cleans and cuts the vegetables. She uses a digital scale to determine the right portion of carrot salad for the food processor, as she is currently on a special diet. At some point Peter is distracted by the television, walking in and out of the kitchen to have a glance at the live soccer game. Most of the time they like to enjoy their meals in the living room in front of the television. However today they are eating at the dinner table, just across from the counter in their eat-in kitchen. Both Amanda and Peter are university students, getting ready to enter work-life. The dinner table often doubles as Peter's office, where he spreads out his laptop and papers. Both set the table together, arrange some food on the plates already at the counter, and some pans are placed on the table. *Bon appetit!*

A thesis on kitchen innovation. The kitchen is a fairly unexplored environment with regards to innovative concepts, compared to, for example, ambient intelligence or pervasive computing. The kitchen appears to be a special environment that has its very own nature, though it is of course part of the

Capture relevant material. Already upfront at the stage of planning the user studies, it is helpful to think in advance if portraits will be made. The combination of observations with interviews is usually good.

Introduce context. Make sure that the audience is aware of where the user is situated, and who is who in the picture. Otherwise it may be difficult to understand how a close-up in the portrait fits into the whole.

Show the person – especially the face. The person's face is the area that people usually observe when they attempt to figure out who the person is.

domestic context. I regard kitchens as a potential host for innovative design ideas, and chose this topic for my graduate thesis in IT Product Design at the University of Southern Denmark. To investigate the potential I conducted kitchen studies to get to know users in their environments.

I conducted studies in four different households, one in Denmark, one in Germany and two in the Netherlands, all with occupants of different ages. This allowed a glance at aspects in and around the kitchen that would possibly inspire innovative design ideas. The main purpose of the kitchen studies was to explore the kitchen context in general, the cooking process and tools involved, the use of appliances in general, favourite interactions, tasks, roles and values. The studies served as a source for scenario creation in the ensuing design process.

Videotaping how people cook. Since I was interested in how people use their kitchens, how they do something and with which objects, I looked for a time of day to visit when most activity would go on – before and during dinnertime. In Denmark and the Netherlands this was in the late evening, in Germany around noon. The kitchen visits lasted between two and four hours, depending on the kind of meal and the social contact. During each visit I tried to keep the same structure of activities, but be flexible to the users' timing and level of involvement. Each visit included the two main parts, observation of the cooking experience and a conversation around props, addressing different topics of interest. After a "warm-up" chat, I tried to stay in the background, videotaping the cooking activities – stepping away from being a guest who demands attention. Now

Let the person tell (and act) the story. Rather than creating voice-overs with a separate narrator, use the original materials from the field. This better grounds the presentation to what really exists in the users' world and increases the credibility of the presentation.

Go slow. It takes time for an audience to conceive things. The editing rhythm for a portrait is usually rather slow. When superimposing texts onto a picture make sure there is enough time to read them. Read them out aloud to find the right time.

Cut meaningfully. Cuts can carry as much meaning as the image itself. As the result of a cut, a clip is seen in the light of another, which may provoke strong associations on behalf of the audience. Think about what

and then I would ask questions, if things were unclear or seemed particularly interesting to me. After dinner the household sat down with me around the table, and we had a talk about their kitchen use with the help of some visual aids that I brought along to document together with the families on-site:

Kitchen layout plan. We reflected on the cooking activity that had just passed, through talking and quickly sketching the environment, the position of things, people and the activities going on. I also made sure to capture some details on paper about all people in each household.

Kitchen appliances on post-its. We further listed all appliances in the kitchen with the year of purchase. Rating them on a scale (frequency of use from "every day" to "never") helped to discuss what they were used for, and how often. This was interesting for getting a deeper understanding of the hidden objects in the kitchen, considering that the one cooking session could not give a complete picture of all appliances that are normally used and the kinds of relationships people have with them. I also looked for the user's interaction with a favourite appliance and the reasoning behind it, to discuss valued qualities.

People and activity plan. To discuss the different roles that the people of a household embody, the users filled in an activity scale describing who takes which actions in the kitchen, which nurtured a deeper discussion.

Screen paper mock-up. Finishing the user study, I briefly introduced the concept of ambient intelligence to the users and encouraged them to reflect on how they could see this incorporated in their kitchen. I tried to trigger their imagination with a plain A3 paper sheet as a screen, which could do anything they would like it to and be placed wherever they wanted it.

you want to say about the person when you connect two clips together,
and watch how the result works. Achieving effective cuts calls for practice but is key to creating strong video portraits.

Let the person explain. Editing software allows adding the voice of the
   person, *e.g.* from interviews, on a separate soundtrack. The person's
   explanation in the background may lend a whole new meaning to the
   activity shown.

Avoid adding music and use special effects sparingly. Sounds dramatically affect the experience that a video conveys. By adding background music
   or sound effects it is too easy to guide interpretation onto the wrong
   track. What if the person in the picture is not sad despite appearing

Video clips, collage and portrait. Having completed the field studies I started by selecting 60 video clips of scenes that appeared interesting – looking
for shifts in action, roles and other possibly interesting aspects. The video
material was used for several purposes and in several formats.

Design potential for explorative concept: First we developed themes for
design potentials, through the video card game. This allowed us to further
develop a design concept based on the idea of sharing and communication
to support social interaction in and around kitchens.

Inspiration collage: Based on these clips I made a collage to gather different
impressions of all four households. The collage was to inspire a group of design students in the analysis of interaction styles in kitchen history and based
on which they would develop concepts for social microwaves. The structure
of the collages was arranged according to the different families, while the clips
chosen to make the collage were focused on the socially interesting aspects.

Cooking portrait: To engage reflection on the microwave concepts developed earlier, I edited a collage that showed a condensed, yet comprehensive,
impression of one family's cooking process. This was then used in a workshop set-up to first mark interesting aspects of roles, space, *etc.* based on a
"real" family, then to look at some of the proposed social microwave concepts
and evaluate them on how socially inspiring they would actually be in a real
use context (based on the portrait earlier shown: "How would their everyday
cooking change through having this microwave at home?"). To do so, the participants of the workshop developed puppet scenarios, where they acted out
the future use of the microwaves in the specific family from the portrait. ∎

to be? Let the audience read the video and make their interpretation through material that remains as loyal to life as possible.

Compared to video collages and stories, portraits differ as to how they foster designers' learning. The making of a video portrait is usually the phase where the most important learning about empathy takes place. During editing, designers need to develop a sense of what they want to "say" through the portrait. Discovering and expressing the essence of someone calls for a sensitive reading of the fleeting hints in the superficial details of the video footage. An engaging and incisive portrait of this understanding also presumes a dialogue between the video material and the intentions of the designers.

A well-crafted portrait both inspires and informs designers about what is valuable to users. Moreover, a portrait helps when designers evaluate designs. For example, a portrait of an ecological cook helps one to see if the

▶ Case story: Freeride attitudes
Salu Ylirisku, University of Art and Design Helsinki

"Let's do one tourist 360," the skier explains while he records the scene at the top of the mountain, where he has climbed with the group of freeride skiers. He self-records the activities, and displays great enthusiasm to capture while the others are climbing. He also captures some smooth pans, like a documentary film-maker, of the group's experiences on the steep ridge.

The Luotain project (introduced in Chapter 2) and the Freeride case had already provided us with some background ideas on the kinds of attitudes the skiers have towards their sports culture. One of the ideas that became salient in the probes, and even earlier when we conducted the initial literature review at the beginning of the project, was that photography and video-making play a significant role in the various freeride skiing communities across the planet. This was one of the "attitudes" that we discovered anew when we were hiking with six freeride skiers in Lyngen, north Norway.

During the four days in Lyngen that we spent with the skiers we captured in total some ten hours of video material. The end result of the case was planned to be a hypertext presentation on a CD-ROM that would contain proposed idea for a new kitchen product fits the person's ideology and way of life. Such a video may also gain personal value for the people involved, beyond the immediate intentions to drive design.

Editing a portrait may require a day or two, but it can be done in a significantly shorter time as well. For example, in the case "Ageing future" that was presented in Chapter 2, the editing of a portrait of a schoolhouse caretaker took only two hours, including the translation into English. This was largely due to adequate handwritten indexing (the time codes on the video) in the field. It helped to pinpoint the right spots quickly. The fact that the video was edited during the same day as the observation, and that the editor was skilled in using the particular software, also quickened the editing process. Creating a portrait can thus be a fast process, when it is prepared well and the material is authored to support this already at the user site. However, the most important should not be the speed but the learning that the portrait editing facilitates. ■

Video example **Freeride attitude** 0'58"

texts, photographs and video clips to explain an understanding of the sports culture called "freeride skiing". The video clips were initially aimed to provide a living picture of what the activity is like. However, the review of the video materials provided insights to a greater potential of the material. It could as well be utilised to delineate the various attitudes of the skiers.

We had gained direct feedback from the skiers to refine our ideas of the different freeride attitudes that we had initially identified, and these seemed to be

● Method: Video Collages

"Defrost-
ing?!
Could this
be under-
stood as
such?"

A video collage is a presentation that combines a collection of video clips according to a thematic principle. When apparently disconnected video clips become associated, and displayed one after another, they develop a new meaning for the audience. This meaning may be explicitly articulated in the name "collage". Sergei Eisenstein, a Russian film theorist, called this effect *montage*. According to Eisenstein (1942, pp. 4–5; in Leyda, 1970, p. 14):

> ...*two film pieces of any kind, placed together, inevitably combine into a new concept, a new quality, arising out of that juxtaposition.*

Seeing things from a new perspective is crucial to finding radically new opportunities for design, and video collages are helpful in this. A video collage may re-

visible in almost all of the studied skiers on the video as well. Hence, we had clear ideas about these attitudes, which helped us to choose materials from the videotapes. The new editing software also had a nice feature in that it was capable of zooming into photos, making them look very good in the video – this invited us to utilise the photographs from the probes kits in the videos.

One of the reasons for focussing on these ideas that we here call "attitudes" of the skiers was the fact that freeride skiing was a rather ambiguous topic to study for product design purposes. Freeride skiing carries an "open sports culture" meaning in that it does not have particular rules, specific equipment, or defined places for the activity. It seemed to be in the people and in their ways of responding to their friends and environments. This, we felt, was nicely captured under the topic of "freeride attitudes", and it also seemed to be present throughout the video material. Thereafter, the editing of the attitude videos was a rather easy task.

The final results, the attitude videos, were each approximately seven minutes long. They had the title, for example "the photographer", visible on the top edge of the screen to help us to remember what the clip was trying to say. Such a technique was not very subtle, but for the purpose, it was considered adequate to make the point. ■

combine real-world activities on video into a radical provocation of new thought.
A new perspective may appear in a discovered analogy between activities and
ideas or in the contrast – how things are just the opposite in different places.

Consider, for example, the collage in the case "Conceptual door", where
little boys are playing with dinosaurs and roar aloud (see the case story later
in this chapter). Seen in isolation it is just a clip about boys playing with di-
nosaur figures. However, when it is juxtaposed with a clip that presents girls
playing with abstract soft pieces of foam while negotiating intensively about
what the pieces mean, grounds are laid for a hectic debate on the meaning of
the differences. What is the role of the appearance of the toys in children's
communication? Do these video clips only tell us something about the differences between the playing styles of girls and boys?

Titling dramatically affects interpretation, whereby the title may provide
a perspective for discovering new design opportunities. For example, a video
collage of jumping people might not be interesting without the title: "Can
humans fly?" Such a collage may turn the design team's focus onto aspects
like "taking off" and "landing", which may influence new sports shoes to
be designed. Hence, a video collage can be understood as juxtaposition, not
only of different video clips, but also of videos and a meaningful label.

The video card game is an ideal primer for the making of a video collage. In the video card game, as mentioned, video clips are grouped into
thematic groups according to a discovered relationship between the clips. A
bit of combining and quick editing can yield an already-titled collage. The
systematic creation of collages is likely to encourage working on material
that is most relevant to the project's focus.

People perceive things differently, whereby it may occur that the original
idea of grouping the clips fades into the background as discussion on the
content is triggered. This is similar to what happens in a video card game,
when activities are rendered into words. Some things become promoted and
some fall into the background.

An effective process for creating a video collage

▸ *Identify the video clips for the collage.* A video card game is a very good
  primer for this.
▸ *Edit the collage.* This may be done even with the "rec-pause" method
  with two interconnected video cameras/devices.
▸ *Display the collage and discuss it.* The discussion is often the most valuable thing in work with video collages.

"Video Action Walls" is another way to craft video collages. This was developed by Buur, Jensen and Djajadiningrad (2004). Dedicated software enables utilising video clips as "living sticker notes" and putting labels on groups of these. This makes the grouping of video material akin to the activity that is usually called "affinity diagramming".

Video collages were successfully utilised in the case "Conceptual door" to discover a new design opportunity for a children's communication tool. The case highlights the value of the video card game and video collage mediated process for perceiving the new opportunity. The attitude video presented in the case "Freeride attitudes" is a mixture of a portrait and a collage. It was created with editing software and it utilised photographs from a probes self-reporting study. It helped to focus on the individual ways to relate to the sports culture of freeride skiing. ■

Co-editing

Interpretation is a fundamental activity in designing. It is essentially about relating the discoveries on video to other discoveries, to earlier experiences, to people's memories, to the organisation's intentions, and to technologies available. Such a multi-faceted endeavour thrives on multi-disciplinary teamwork, as the backgrounds, aptitudes and biases of different members come together to form shared understandings. Video can support collaborative sensemaking by rendering observations open to scrutiny. The term *sensemaking* underlines that interpretation in design is a creative process not just of finding meaning, but of constructing understanding. A video-mediated sensemaking process is an opportunity to learn about oneself and about others. – Oh, is it possible to see it that way? – Can I understand how it could be seen that way? – Does it move us ahead with our project?

When working with video there is a unique opportunity to express interpretations of field observations in the media itself. The act of collaborative interpreting turns into *co-editing* of video stories, video portraits, and video collages. Video artefacts are instantiations of new meaning. They constitute a move towards the next design stage: to the activity of creating. Creating with video is the topic of the next chapter.

- I'm making a door.

▶ Case story: The conceptual door

Johan Karlsson, HDK School of Design and Crafts at Göteborg University

"You are a turtle. I am a squirrel. I am mom." The children are negotiating who they are in the video. The whole design team laughs aloud – including the teachers who ought to be the critical tutors guiding our progress – as we watch the video to study how children communicate. When we entered the realm of children's fantasy we were faced with the incredible creativity of the tiny inventors.

The User Inspired Design project focused on designing product concepts based on a very open brief.[†] We were given the task of designing a new concept, which used the idea of "door" in some way, and which was based on the study of a relevant user group of our choice. We focused on children in daycare with the intention to understand "door" as a concept related to understanding each other – as the discovery of "a door" to another's mind.

After the visits we prepared for a video card game and picked out clips that contained various ways how children communicated. We made the clips rather short and the cards big to aid the writing of the observations on the card. When the grouping of the cards into themes began we spent

The User Inspired Design course was held at the University of Art and Design Helsinki, and lasted for thirteen weeks during autumn 2005. The team members were Pia Salmi and Yun Yegal (MA

[†] students), Sara Estlander (student of Cognitive Science) and Johan Karlsson (exchange student at the university from Göteborg University of Craft and Design).

Initial themes

- Copying and intensifying action
- Hanging out together
- Helping each other
- Playing individually together
- Coping action without intensifying it
- Joining a game
- Collaboration around the bricks within group activity
- Deciding together about what to do
- Playful arguing and fighting
- Intensive silent working together
- Setting the rules for others
- Taking initiative to behave different than others
- Not cooperating
- Enthusiastically waiting for ones turn
- Building subgroups spontaneously
- Doing something individually
- Focusing on one thing while loosing track of other things
- Interaction with the camera or the camera person
- Teachers guiding action before an event
- Adults resolving conflicts
- Collecting the children and calming them down for rituals
- Singing and acting at the same time
- Communication by acting

Regrouping

Identified themes

Imitating
- Taking after each other
- Often intensifying the action
- A way of creating new games

Playing alone in a group
- Doing their own thing in company
- Only reacting to what is there, not "meeting each other as humans"

Deciding together
- Talking and acting before and during play in order to try to reach agreement

Helping each other
- Helping someone else verbally by giving directions or physically by showing

Identified playing roles

Director
- Lays down rules, accepts or rejects others' ideas, gives orders

Initiative-taker
- Presents ideas

Follower
- Goes along with others

On one's own
- Plays alone or watches other
- Sometimes in the middle of someone else's game

Figure 3.8
Identifying themes and roles from daycare material

much effort in negotiating the titles for the themes. We wanted to make them as concrete and inspiring as possible.

Quite soon we had a list of themes with names describing what happened in the clips. At the end of the session, the 80 cards had been divided into 24 groups that all described a particular event or action (see Figure 3.8). During the game we stamped the cards onto A3 sheets and put them on a wall to see the overall structure. We re-organised the sheets for a moment, and the themes were grouped into five major groups. Some leftover themes did not easily group with any of the categories that we had, but time did not allow us to discuss further in the game session.

After the video card game our student team held a meeting. We put the sheets back on the wall and re-organised the whole. We made the five major groups more clear with more precise titles. After intensive discussion and brief review of the written notes from the kindergarten visits, we ended up in four theme groups and a group about the playing roles of the children. In this re-grouping we kept in mind our goal to design something related to the idea of a "door to another's mind". In this phase we did not quite know what else that would mean, aside from focusing on the different ways children communicate.

Based on the identified themes and roles we edited video collages that we presented during a lecture. The making of these affected the discovery of our final design idea. In the collage that we named "imitating and intensifying action" one clip showed a group of boys playing with dinosaurs. Sitting in a ring, a boy burst out singing, "dino dino dino!" and soon the two others copied and reinforced this. Similar examples of reinforcement and exaggerations of behaviour could be seen in a number of the clips, among boys and girls, inside the centre and outside in the yard. In contrast to copying action, we had clips where children were negotiating instead of merely copying each others' behaviour. The collage "deciding together" showed a group of girls playing with boxes. They were busy discussing what they where doing, who was what, and what the boxes were. Similar discussion happened in the video clips where children were modelling children's clay. These collages enabled us to compare the situations and environments that triggered these different behaviours.

The main difference that we found was that the children, both girls and boys, discussed more in the video clips when they were playing with abstract forms. The boxes and clay raised questions and provoked talk, as opposed to the figurative toys, such as the dinosaurs, that either caused the children to shout aloud or made them lead their own play individually. As an overall result, the video card game helped us to realise the relevant themes in children's communication that enabled us to build the product concept, which eventually proved to be quite interesting for the children. The concept and its evaluation with children are presented in the Let's Playnt! story in Chapter 5. ∎

4

We are running on a minimum level

Envisioning the future

JOHN DEWEY

"To be playful and serious at the same time is possible,
and it defines the ideal mental condition.
Absence of dogmatism and prejudice,
presence of intellectual curiosity and flexibility,
are manifest in the free play of mind upon a topic."

4

Envisioning the future

The study of users focuses on current use practices, *i.e.* it tends to explain history, where design needs to move forward and understand *changes* in practice. The difficulty in making this move is to see how change helps to develop current situations into preferred ones. This is where video scenarios about possible futures are valuable. Scenarios are utilised in design, engineering and marketing to inspire and develop ideas, to test ideas against conceived reality, to discern requirements, to establish a common ground and to communicate ideas. Use scenarios constitute one of the core tools of the user-centred design process.

The discipline of scenario design borrows heavily from theatre to enable designers to move beyond merely discussing futures, to *enacting* possible practices with proposed designs, be it in the design studio or out in the real environment in collaboration with users. When stories are acted out, the sensual engagement allows situations to be understood on the level of both bodily and social performance. Therefore theatre concepts play a central role in this chapter.

At the same time, the process of play-acting and movie-making in itself establishes a playful and creative atmosphere that bonds team members, facilitates user collaboration, and bridges disciplines. Video recording the scenarios provides both an excuse for "acting funnily" – "we act for the camera" – and an incentive for reflection, when recordings are played back for evaluation. As with the video stories, portraits and collages of the previous chapter, enacted *video scenarios* constitute important moves in the design process. However, it is not video *per se* that creates the impact, but rather how its role is constructed. This chapter outlines a process of building an effective role for video in the acts of envisioning the future.

This chapter discusses three core concepts for scenario design, namely "improvising", "ethnography of the future", and "directing". "Improvising" highlights the *courage to explore freely*. "Ethnography of the future" outlines the importance of *building an understanding* of the design opportunities upon the observed reality. "Directing" promotes the *controlled and conscious planning* of images of the future. Through entering the unforeseen and linking ideas and impressions together to build new aims, a design organisation may engage in a reflective discovery of new potential.

The future as theatre

Interactive technologies bring an increased complexity into everyday products. No longer can designers simply focus on the interaction between one product and one user; products have a much more profound influence on people's daily practice and indeed their lives. Computing technologies are not merely employed to accomplish practical goals; they provide pleasure and serve as cultural symbols to express social identity. Hallnäs and Redström (2006) suggest that computing technology has become a form of design material, albeit with the capacity to complicate people's *interactions* in unforeseen ways. As a consequence designers need new ways of imagining, exploring and trying out completely new "realities" of environments, social relationships, and practices.

Theatre with its ability to create settings that allow actors and audience to explore human relationships has much to offer design. The staging of plays in an imagined reality allows freedom from the constraints of our current reality. Yet theatre bridges imagination with the firm ground of sensual and analytical knowledge of what the reality is like. It may even utilise physical elements of our everyday surroundings and stage the plays in the environments where people live. Theatre nurtures images of current reality and of the future that promote an increased awareness about what could be desired.

To see oneself act
Augusto Boal, one of the revolutionary reformers of theatre during the 20th century, begins his book *Rainbow of Desires* (1995, p. 13) thus:

Theatre – or theatricality – is this capacity, this human property which
allows man to observe himself in action, in activity. The self-knowledge
thus acquired allows him to be the subject (the one who observes) of an-
other subject (the one who acts). It allows him to imagine variations of
his action, to study alternatives. Man can see himself in the act of seeing,
in the act of acting, in the act of feeling, the act of thinking. Feel himself
feeling, think himself thinking.

Boal devoted his career to promoting a high involvement of people, especially
"ordinary" people – the oppressed – in designing theatre plays and acting
in them. Boal's central focus was on the theatrical methods that help to increase people's awareness of themselves and of the affairs in which they are
involved. This links to the interest in user-centred design: it is the awareness
of everyday reality that grounds the arguments for the value of products. It
is this everyday reality as perceived by (ordinary) people that Boal's theatrical methods address.

Boal contends that increased awareness has an inevitable impact on the conditions of everyday life. Especially, increased awareness helps transform the conditions of life into a more desired direction – which is also the essential aspiration of design. What will this mean for us? Reflective theatre – as defined by Boal
– and user-centred design are surprisingly similar affairs. Boal's ideas are rooted
in the power of theatre to provoke people to perceive things in new ways.

Let us consider a concrete example in the form of a play. An actor raises
her hand as if to take an imagined teacup from the rack. She puts the cup,
a plate, and a napkin on the tray, and pushes it along the invisible counter
towards the cashier. The actor has nothing but her body and the concept of
buying tea to work with. She relies on the audience's skill in constructing
what happens as an image in their minds from the cues of her movements.
Boal's "Image Theatre" works by displaying people's conceptions of things
through bodies, gestures, orientation, voices, and people's locations.

At the other end of the scale of realism is Boal's "Invisible Theatre". Here,
scenes are played in a public place without the audience knowing that they
are actually witnessing a theatre play rather than real action. Invisible Theatre provokes people to think about the burning questions in society by involving them in situations where these issues are highlighted. This conveys
a highly realistic experience of the incident in question.

Like theatre plays, video scenarios can be acted out minimally in an
empty studio or in the users' native environment with all the tools, people,

and even real activities unfolding in parallel. The world of theatre (or video scenarios) is, to an extent, a world of symbols or symbolic acts that evoke interpretation. Regardless of whether the plays are enacted by actors, by users in their native settings, or by designers in a studio, the scenes will involve both the actual, physical interaction and subjective perceptions. In this way, when the acting is *about* something, it enables us to escape the constraints of physical reality and delve into imagination.

A theatrical approach to design fosters reflection on everyday life as conceived by people in high detail. Theatrical techniques enable us to construct complete and sensitive visions of the future embedded in concrete situations. Through the elaboration of situations, theatre presents issues in relation to people, activities and environments, and easily accommodates both the past and the future. It works with the images that people have in their minds as well as with material reality.

Theatre carries three fundamental characteristics as to why it functions as a practical tool for envisioning future scenarios in product design:

▶ *Firstly*, theatre has the *capacity to contextualise* things in a most detailed manner. Theatre plays can be acted in real user environments and with real tools. The actors may be the potential future users themselves, acting even in-between their real life tasks. Such capacity to contextualise brings new relationships and meanings to light.

▶ *Secondly*, theatre facilitates *immersion* into the world that a theatre play depicts. This immersion evokes personal responses and helps to bring to the surface individualised knowledge, images and attitudes that people carry within.

▶ *Thirdly*, the *means of expression* in theatre, such as speech, mime, music, and dance, and any presentational media imaginable, bring specific qualities that are held to be the exact character of these means. These expressions function beyond being symbols. This is much more than what designers can bring to stories through a rational envisioning. A theatrical approach to video scenarios can enable a revelation of relationships that would otherwise remain silent, and hence, be ineffective for design.

The making of video scenarios has inherited much from theatre. However, the presence of the video camera affects the nature of this "theatre" in respect to several aspects. Firstly, video fosters an even deeper reflection on the plays due to the caft that the plays can be reviewed later. Secondly, video allows editing of a presented play into a new form. This may affect both the acting as well as the reflection upon it. For example, if we consider movie-making as a kind of theatre, where the acting becomes coordinated by the use of the video camera, the activity is completely coloured by the collaborative intention of creating good video footage.

Video scenarios, like theatre, build on *acting*. Acting imports everything that people are to these plays: their appearance, ways of moving and responding, expectations and attitudes, voice and other personal characteristics. Acting with others will also bring the culturally attuned, and unwritten, rules of social cooperation into the play. Building the story together with users is a means to bring the experience of the people and their memories of relevant real-life situations into the story. Moreover, as the users are deeply involved in the practice of their domain, they carry a broader understanding of the practice to the scenario authoring situation through their professional identity. To put it simply, the collaborative authoring of plays enables discovery of what people know about a topic and gives them an opportunity to contribute with their knowledge.† Eva Brandt (2006, p. 64) outlined its importance thus:

> ...*designers need a framework that helps organising participation in such a way that the various competences present in an event can be utilized, that everyone can make design moves and be part of exploring and negotiating views in order to create common images of possible futures and the prospective design work.*

† The notion of knowledge is a disputed issue. Here knowledge is understood in the way Wenger (1998) outlined it, as "a matter of competence in valued enterprises".

Giulio Jacucci (2004), who studied performances in the design of mobile applications, argued that the "traditional" way of creating static and task-oriented textual scenarios was too limited. Such scenarios were not sensitive enough for the modelling of interactions for mobile applications. Real-life situations are coloured by the contingencies of ephemeral, unique events, personal means of self-expression, and communication. These details escape generalised descriptions – but influence people's everyday interactions dramatically.

Users may be involved in the authoring of scenarios in a variety of roles. They may turn into scriptwriters, actors, cameramen, directors, and even editors. Especially when users act themselves, they are often able contribute wonderful details of their characteristic ways to do things. Moreover, assigning the users a "role" of a specific kind provides them with a rather easy-to-understand way to orient towards the design of the story. Every one of us can imagine what a director does, even set ourselves into that role and start directing a roleplay on film. Thus it helps to frame design within the playful and imaginative world of theatre and moviemaking.

Design as theatre

In her book *Computers as Theatre*, Brenda Laurel (1993) counters the rational design process of computer science by suggesting a new theatre approach. The approach promotes human experience over technical performance. Laurel was struck by people's reactions to computer systems. People were often distracted by the need to figure out what the system was doing during attempts to negotiate through the interface. Computer systems were built in a manner that divided the whole into two parts: the *functionality* and the *interface*. The interface was often created after the functionality, whereby the design was largely driven by the technical rationality of the functions. By proposing the theatre approach, Laurel attempted to raise the understanding of how people experience software systems. She tried to move the focus from what is possible to what is desirable. Laurel's shift reflects the broader transition of focus from usability towards user experience during the 1990s.

User experience, however, proved to be an extremely difficult to define topic. A consensus on what constitutes user experience does not exist. The researchers, designers, psychologists and marketers who have approached the issue have merely framed the topic from their own relative background and interests.[†] One of the most insightful studies into user experience is that of Battarbee (2004). She reviews a number of frameworks and approaches to defining user experience and develops an understanding of user experience as a phenomenon that is fundamentally social and constructed. Experiences are created in social interaction in real-life situations, and the interpretations of these develop over time. Remarkable in this observation is the essential role of *situations* as well as people's *personal interpretations* of these.

Where Laurel and Battarbee attempt to better understand the meaning of situations for people, Boal's remark about the simultaneous "subjective" and "objective" character of theatre promotes the importance of understanding *both* the social construction of the meaning of situations as well as the material interaction in them. Theatrical methods allow designs to be placed into a conceived use situation, and both the material influence and the impact on people's interpretations of the situations are explored at the same time.

[†] See, for example, Rhea (1992), The Experience Lifecycle; Pine & Gilmore (1999), Experiences of Various Kinds; Kankainen (2003), Experience as Motivated Action *In Situ*; and Garrett (2002), Elements of User Experience on the Web

A video scenario conveys the causality of actions in a concrete way. When incidents follow one another in a play, the audience makes causal inferences. One thing leads to another, and provokes a response. Previous events form the backdrop against which the current event is evaluated. This forms the basis for judging how new ideas provide people with value, and is the reason why macro scenarios, like the ones created in marketing to forecast changes in the economic environment, are not very useful for design. Design requires an active engagement with the micro-level relationships between people and technology: the dynamics and detail of *how* a product functions in social interaction, *how* it creates value. Addressing the details of interaction is necessary for understanding how a design fits and is adapted in people's practices. The timely and concrete nature of video scenarios forces the design team to think of the dynamic nature of interaction between people, the product and the physical environment.

Acting out departs conceptually from action by *representing, or being about,* something. Action, such as what people usually do in their environments, *is* something. When ethnographers try to read the conceptual patterns written in human behaviour (like Geertz [1973] guides us to see), the acting out also underpins a message carved into behaviour. However, this message can be explored and creatively manipulated through theatrical plays. Moreover, with video scenarios, this message is mostly about the intended meaning and value of the new product ideas.

● Method: Video Brainstorming

Video brainstorming was originally developed by Mackay and Fayard (1999) to explore and capture design ideas in an interactive manner. The method aims to generate ideas on how people interact with technology and elaborate some of these through acting them out. Capturing ideas on video has a profound effect on the process of ideation and on the way in which the selected ideas are elaborated. The ideation process shifts from a discussion around the table into a lively and playful enthusiasm around the video camera. Who shall operate the camera? Who shall be the actor? What shall she say and do? The situation encourages participants to elaborate their ideas from the point of view of human interaction, and thus new social issues may be identified when people collaboratively "run through" the ideas. This may provide new inspiration as well as constraints on the ideas.

"How would your idea be used?"

Video brainstorming follows a three-phase pattern

- ▶ *Step 1:* Create ideas in an ordinary brainstorming fashion and list them on a flip sheet.
- ▶ *Step 2:* Choose the most interesting ideas. The participants should go through the entire list of ideas before they cast their votes for the best ones.
- ▶ *Step 3:* Act out the most promising ideas in an improvised way while recording them on video.

The camera is used for capturing the ideas. This means that everything required to later make the idea understandable to other people needs to be recorded by the camera. The actors collaborate closely with the camera operator to negotiate which parts of the interface and interaction should go on tape, and where the camera should be positioned and pointed towards. The team simply uses the record button to start and stop the shooting from different locations to coarsely edit the video in the camera. A facility for an instant review of the video helps to ensure that the ideas are expressed with enough detail. The ideation continues while creating the video recordings. Ideas may build on each other, and, for example, turn into a series of subsequent interactions with a product.

Mackay recommends the added discipline of introducing each new idea with a handwritten board including the title, date, and authors' names, so the resulting recording becomes a collage of idea scenarios. This small trick of naming helps the design teams to focus on the essential in each idea.

The key advantage of video brainstorming is the speed-detail ratio of the result. Although acted out with rough materials, such as pens and paper, the ideas are expressed on a surprisingly high level of detail regarding the interaction. Achieving the same amount of detail with other methods is likely to require much more time.

The following case story "Phoning a deaf person" presents a variation of the video brainstorming method. The story shows how the ideation of the relevant design ideas is first grounded in the real-life experiences of the workshop participants. Moreover, the fact that the real users participated in the brainstorming proved to be a valuable asset to the presented design project. The sketching of design ideas in quickly crafted plays resulted in new, detailed and highly engaging mock-up ideas described in the contextualised language of human interaction. ■

▶ Case story: Phoning a deaf person

Bo Westerlund, Sinna Lindquist, KTH *Stockholm*

During a workshop on video-mediated telecommunication at the lab, one of the participants tells us a story illustrating constraints in his everyday life: Ragnar wants to get rid of his old sofa. One way is to put a note on a public billboard saying: "Sofa for sale! Tel. 0735007076." However, Ragnar is deaf. Anyone can phone him, but he cannot answer.

This story is perfect to build upon. It is concrete, has a clear aim and a defined problem, with enough complexity and visual aspects. "How can Ragnar sell his sofa?" This question is posed to the workshop participants, and the group has numerous ideas for solutions. Someone suggests:

– We can connect the two parties with a sign language interpreter through mobile video telephones.

– OK, great. Let's shoot this. Who will phone Ragnar and who will buy the sofa?

Video
workshop
**Deaf people
and phoning**
8'35"

Video
examples
**Phoning a
deaf person**
4'07"

Improvisation

In improvisation people use their entire body to explore and express ideas. Keith Johnstone (1987), a virtuoso reflective improviser and theatre educator, wrote that improvisation is like walking backwards into the future: the walker may not know what lies behind him (in the direction he is actually heading) but knows the path from whence he came. Improvisation is a way of engaging people in creating the new. It is involving, multi-faceted, holistic, social, natural (you may read: easy) and fun.

Johnstone (1987) suggests that improvisation is essentially a means of

This story is the starting point for a video prototype: an illustrative, collaboratively made artefact, showing future use of future products and services.

A workshop on technology and disabilities was held in KTH, the Royal Institute of Technology in Stockholm in December 2004. Twenty people met at the lab at CID, the Centre for User Oriented IT Design for five hours of hard work. The workshop was one in a series of workshops exploring how technology could be of help in everyday life for people with disabilities. Most of the workshops were done in collaboration with HI (the Swedish Handicap Institute). In the first workshops we worked together with people with cognitive disabilities. Another workshop had participants with several disabilities, physical as well as cognitive, and was focused on electronic payments and ATMS.

The aim in this workshop was to explore the design space for future mobile video telephony. This was done with the help of deaf persons using sign language as well as participants from mobile phone manufacturers and service providers. The streaming of video mobile phones means a revolution for the deaf community. They can now talk to each other and their relatives at a distance in their own language.

We had learned about the video prototyping methodology from Wendy Mackay when we were working together in the participatory design project InterLiving† some years earlier. In that project, the video prototypes helped us to construct understanding and formulate ideas together, both researchers and user participants.

InterLiving's †
Web site:
http://inter-
living.kth.se/

breaking routines, and this is necessary for discovering radically new ideas.
He observed that people at some phase in their lives appear to lose their
creative childhood imagination. A strong sense of right and wrong ways of
thinking learned in school effectively blocks creativity. People hesitate to as-
sociate freely and to express ideas openly. Free improvisation presupposes
that one can let go of control, and most people will object to this.

Why is it that we want to be in control? What are we afraid of? In the
workplace we nurture an image of ourselves as sensible members of our
working community. This is visible in how we dress, how we behave in a
group, how we speak, and what we speak about. When control is lost, so

The workshop began by grounding ideas in the lives of the participants. The
participants were encouraged to tell stories about situations they had experienced as important and meaningful. They could describe both problematic and
pleasurable events. Instead of general descriptions that lack detail, we asked
the participants to share experiences from actual situations and also make the
context comprehensible to all the participants.

These stories were followed by group discussions to search for possible solutions for the problems identified in the stories. These explorations resulted in
articulation of new ideas in the form of scenarios, which were written or drawn
as storyboards. The scenarios facilitated building a common understanding of
the relevant issues, and they were applied as the basis for authoring the video
prototypes.

To illustrate characters and ideas in the scenarios the participants then made
quick-and-dirty prototypes. This was a fun and engaging activity. Before starting
shooting we explained what was going to happen, and what was expected from
the participants. This was done to ensure that everyone was feeling comfortable
at the time of capturing the scenarios. We also emphasised that the objective
was to visualise the ideas, needs and desires – not to make good-looking "movies". Then with a little assistance in shooting, the group acted out the scenarios
with their props.

The last activity in the workshop was the collaborative viewing of and discussing of the scenarios. The watching of the videos both triggered new ideas
and provoked some criticism, for example, about the relevance of the scenarios
to the participants' everyday life.

is the image we are trying to hold on to and trying to project to others. Improvisation may reveal sides of us that we are not willing to share with our workmates. There is hence good reason to be frightened of improvisation, and therefore, improvisation needs to be approached with sensitivity.

Johnstone (1987) emphasises the importance of constructing a situation where people are not made responsible or punished for the things that their imagination creates. Ideas must simply be accepted at face value. Only later can meanings be explored and values assessed. When improvisers are instructed not to be responsible for their ideas, it helps them to overcome some of the barriers that block imagination. However, once improvisation

After the participants had left we went through the workshop and evaluated what was good and what went wrong. We also collected, labelled and archived the different artefacts that were made. This helps greatly when returning to them.

The video prototypes are rough. They are a means for generating and conveying design ideas and developing an understanding of the relevant issues in the participants' lives. The crude format of the video prototypes is purposeful at this stage. Video acting should help make the ideas detailed and clear, which helps convey them to others. Making the prototypes does not require "acting" in the theatrical sense. People are rather playacting as themselves in staged situations. Since all participants collaborate in the making of the video-prototypes, the event leads to shared experiences where the understanding of all stakeholders' views and skills grow.

One important aspect of video prototyping is that at the end of the day you have complete short films illustrating people's everyday contexts, their needs and desires as well as ideas for solutions to problematic situations. You do not need to look through hours of video, analyse, interpret and describe this yourself. Moreover, when the ideas are grounded in people's stories about their real experiences – it is for real! This enables designing for a real situation. When the participants also develop the ideas themselves, act them out and discuss them, this kind of workshop produces highly relevant ideas. ∎

as a skill develops and people gain confidence through practicing, they will
also learn to take responsibility for their ideas.

Especially in product concept design the breaking down of personal barriers to foster creativity is a very sensitive issue. Although Johnstone encourages people to accept all ideas that occur to them and to avoid taking responsibility for what comes out, this may not work (or even make sense) in all cases and with any combination of people. Johnstone suggests techniques for ensuring an unrestricted flow of imagination: counting backwards in the mind while creating a story, for instance. This overloads the cognitive capacity to control, and results in free writing of whatever comes to mind.

Improvised video scenarios are a very powerful means of exploring the design space and developing early ideas, but many people, designers and users alike, do not feel comfortable with acting in front of a video camera. To achieve this, one must strive for a creative and playful situation, where people are encouraged to laugh at themselves and each other. This does not mean to ridicule the situations or the people involved but to overcome the barrier of being too afraid to contribute. For example, Tom Kelley, the IDEO CEO, suggests playful rules as one of the key features of fruitful ideation (Kelley, 2001). Improvisation, like brainstorming, calls for a willingness to cross the border of rationality and enter the realm of wild inspiration. This is not possible in a mood of critical judgement.

An important blocker of imagination is set by the high expectations of the designers. For example, Johnstone (1987) observed that when people try to conceive *original* ideas during improvisation they usually end up with rather mundane and unoriginal ideas. He asserts that when people accept the first thoughts that come to their minds, they will be driven to a more resourceful ground. People are often delighted by the most self-evident ideas. Similarly, story writing can be quite difficult when people strive to author a good story. Johnstone (1987) observed that when people are, instead of writing a story, asked to describe a routine activity and then to destroy it, they do not have any problems. For example, a routine could be walking through a forest. Johnstone uses the term "routine" to refer to an activity that everybody would expect.

This kind of creativity calls for high tolerance of irrelevance. Ideas may at first appear as foolish, insignificant or extremely risky. However, they may gain meaning from the next idea that appears. When absurd ideas are combined, they may accrue a meaning that makes perfect sense.

Constraints enhance imagination

A healthy imagination is always constrained in some respect. The psychologist Rollo May (1975, p. 135), who studied creativity, points to the crucial role that limitations play for creativity:

> *...creativity itself requires limits, for the creative act arises out of the struggle of human beings with and against that which limits them.*

Form (which in May's terms includes also non-material matters) is a fundamental composer of boundaries and structure to a creative act. Limitations are set by our material reality as well as our subjective perceptions. Designers may intentionally adjust these limitations according to purpose. For example, providing a heating installer with a design mock-up effectively focuses the ideation on the features of a product with such a form, weight and size. Similarly the design brief at the outset of a design workshop sets a structure and border for thinking. An effective presentation of key findings of a user study at the beginning of a collaborative video scenario workshop assigns a background, or form, with which to work. Constraints may be expressed as subtle cues, such as choosing the right environment, or strict rules, such as in design games (see, *e.g.* Ehn and Sjögren, 1991).

May (1975) sees limitations as "river banks" that canalise spontaneity. Constraint delineates a border where things may be related, and on which new things may grow. Due to the enabling, rather than limiting, role of constraints in improvisation, it might be a good idea to understand their role as *givens*. They release and focus mental energy on the issues that may change. Brandt and Grunnet (2001) state that:

> *For instance it should be easier to improvise a use situation when having a specific user in mind than just improvising as any user. In this sense restrictions or guidelines give the users or designers something to hold on to from which they have to design.*

Video scenarios offer the design team a range of opportunities for delineating background and structure, and for setting the borders for the scenario building and reflection in the design event. Tools, such as mock-ups, scale models and design games, are applicable. For example, in the case story "Puppets in the kitchen", a video collage was utilised to stage the event of imagining new design opportunities for the kitchen with puppets. The physical setup for the improvisation was built of cardboard and tiny dolls. Video plays a two-fold role in setting the stage for the event: firstly, it presents the use context, and secondly, it provides a "moviemaking" frame for crafting the puppet scenarios.

Another reason for providing various forms to guide the improvisation is the fact that *people cannot spontaneously provide their relevant knowledge.* If knowledge resides in action (Wenger, 1998), it also resides in *interaction* between people and their environments and becomes mediated by the tools people use. Knowing forms a process involving the environment and the people. Donald Norman (1988), for instance, explained how parts of our knowledge are located in the world, and introduced the example about the details of a coin: can you draw the figure on the front of a five-cent coin? Therefore, to enable this knowledge to surface, people need to be provided with sufficient provocative tools that help to generate an understanding of this knowledge. Sanders' (1999) make tools rest on this idea, as do collaborative video scenarios.

● Method: Puppet and Mask Scenarios

In puppet and mask scenarios the actors play-act through representations: they move puppets or talk behind masks. With these techniques, untrained actors do not need to put themselves on the line: they do not need to draw attention to their body or face. This is comforting for shy people in particular. Instead the participants need to project their ideas and attitudes onto the representations, and communicate through movements and speech.

"You are
the intelligent phosphorus
sensor"

Puppet and mask scenarios foster verbalisation: holding a rather static puppet – or mask – in hand, participants are forced to verbalise what their puppet thinks, aims to do, and how it feels. This verbalisation can lead to new understandings and meanings. Although both scenario types work with very cheap materials and little preparation, they have slightly different advantages.

Puppet scenarios provide a good overview of what several actors do simultaneously. The small scale offers a "God's eye" view of a small community: the actors can see other puppets even though they are in another room, another building, or another country. There is little focus on precise interaction with technology; rather, the puppet scenario allows participants to work with overall social relations and general functions and services.

Mask scenarios are to scale, but all masks do not need to be human – technology may also speak and think. In the "Intelligent pump station" case,

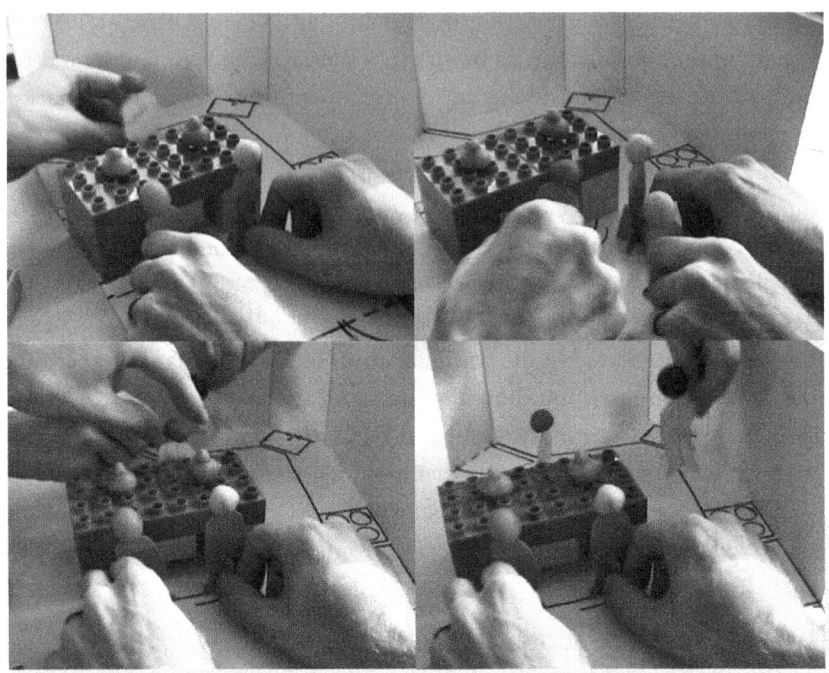

Video scenario
Puppets in the kitchen
0'59"

▶ Case story: Kitchen puppets

Mette Mark Larsen , University of Southern Denmark

The process of making the puppet scenarios seems more beneficial than the scenarios themselves. We really learned a lot about microwaves, says one of the five participants, when we evaluate the outcome of a half-day design workshop that had puppet acting as the main activity.

Social Kitchen (the project was introduced in Chapter 3) was an effort to design kitchen appliances that support social interaction between family members in the kitchen. At the time of this workshop, I had completed studies of four families, and a student design project had helped establish five novel microwave concepts for future kitchens. What I wanted to explore was how "social" these concepts would actually be in a real family. I selected the two most promising: "Tada" – a microwave oven embedded in the dinner table to allow a family to cook together, and "Ladybug" – a round-shaped glass microwave oven attempting to turn cooking into a visual experience for several people.

for instance, designers try out what roles technical components of a future machine system would play. This allows the team to imagine opportunities "as if the components were intelligent".

With puppets and masks a design team can easily enter a constructed reality, whether created from real experiences or imagination. The method uses cheap and available materials, and is a convenient way to simulate complex future situations, which would require extensive work if authored with sophisticated tools such as 3D modelling.

Setting the stage

Create puppets or masks. Puppets can be paper dolls, small toy figures, LEGO bricks, or even bottle corks – basically anything at hand that

To get a shared understanding of the use context as it is without one of the new microwave concepts, the participants watched my video of how cooking in one of the families proceeded when I visited them. After brief introductions to the two new design concepts, I asked the five researchers and graduate students to develop puppet scenarios that would show aspects of how those microwaves could change everyday behaviour in this household. As we went along I recorded the scenarios on video. Each group had a table set aside for playing the scenarios with a small-scale cardboard model of the kitchen space, the original layout of the family's kitchen drawn on the ground, main walls, and some indication of the connected rooms. I also provided each group with a set of cardboard/clay puppets, representing the five family members. The participants were encouraged to rebuild parts of the environment if changes were required in their scenarios. For this purpose I provided a variety of material: Legos, wooden bricks, clay, straws, tin foil and similar articles.

After videotaping the scenarios, the participants in their groups discussed the impact that the concepts had on the household. This then developed into a final discussion on how "social" the concepts really would be, and whether this family realistically would desire to own one of the microwave oven concepts. Here participants also related their own impressions of whether they could see one of the microwaves in their own kitchens. Many relevant and most interesting reflections on the concepts came up in the final discussions, triggered by the puppet scenario acting. ■

enables the designers to think of the "puppet" as being someone or something.

Prepare an environment. Puppets can play on any two-dimensional layout, like the floorplan of a house, a plant, a school. Or a map of a city, a shopping centre, a road map. With user collaboration it may be an advantage to play on an authentic plan that they have brought themselves. If required, simple materials may be used to create walls or furniture. Mask scenarios require a full-scale environment, which can easily be established using cardboard and paper.

Establish a story and assign roles. As with any scenario method there needs to be a story with an aim upon which to improvise. Moreover, the actors need to choose their favourite roles. With users, puppet scenarios are a good way of checking the outcome of observation studies: "We saw you working over here, but where did your colleague call you from?" It is easy for operators in a brewery, for instance, to play out a situation that they have recently experienced. This sets the atmosphere for thinking about changes to the routine.

Explore opportunities. Then the planning of the plot for the play begins. This is often a hilarious activity, as people are quite enthusiastic about the chance to play with toys. There is, of course, a risk that the action gets out of hand and has few results for the project even though the participants are enjoying themselves. Participants are free to suggest their ideas of what might happen and how the plot should develop.

Document on video. Do not work with a fixed camera, move with the action. With puppet scenarios, use low, wide-angle camera positions to get on eye-level with the puppets.

Facilitate the action. The role of the facilitator is to ensure that everybody can contribute to the development of the story. If the team is stuck, the facilitator may ask individuals how they would continue.

Make room for reflection. The discussion after the ideation and scenario acting is where the key lessons are learned. It is invaluable to see what has been produced and reflect on the potential.

Puppet and mask scenarios are typically employed in a phase where the design project seeks to discover radically new ideas. They are fun to make and foster creative collaboration. ∎

![Photo of a person in a striped sweater holding a paper with a drawing of a device, standing in an industrial room with pipes overhead; another person in the background holds a paper in front of their face.]

▶ Case story: The intelligent pump station

Jacob Buur, Danfoss User Centred Design

Mask
scenario
**Intelligent
pump station**
2'48"

– Is everybody ready? OK, camera is rolling!

Ole acts as a process operator. He unlocks the door to our improvised pump station stage, enters the room, and introduces the components one by one:

– Here's the flow meter, measuring the inlet side.

Ole points to Jens, who holds a simple flow meter illustration in front of his face, just like a mask.

– And over here we've got something new, two Evita sensors that measure nitrate and phosphate.

Pernille and Hans are acting as sensors. After having introduced also two pump controllers (Jesper and Lotte) and the automatic valve (Kirsten), Ole leaves the pump station, but the camera stays. Then, the flow meter calls out:

– I can feel there's a bit of rain coming.

– Everything looks quite normal here, says the phosphate sensor, having conferred with the other sensor.

– We're running at a minimum rate, says one of the two pump controllers.

Now a little drama starts unfolding:

– There's more water flow coming now, maybe it's raining

– Our sensor level is really coming up, this feels like the "first flow" situation of heavy rain.

– We'd better pump faster, says the pump controller, and they start shuffling their feet to indicate something is moving.

– Much more water is coming now, more flow than the plant will be able to handle.

– It looks quite clean now, should we take action and divert the flow?

The components keep negotiating for another few minutes with the stress level building up. The intelligent components discuss what is happening and if they should route the heavy wastewater coming in directly to the sea, as it is obviously clean rain water, or if they should keep pumping it into the wastewater plant with the risk of running beyond capacity. They send an e-mail to the operator, but getting no answer, in the end they decide to de-route the flow to prevent overflow.

The water vision project (introduced in the case "Plant operators") where the "Intelligent pump station scenario" was created looked at new technology opportunities for, *e.g.* wastewater treatment plants. One of the main questions that the team struggled with was this: How will future sensor and controller networks influence the work of process operators? Consequently, how shall their user interfaces be designed in the future? The major shortcoming of most industrial components today is the link to the human practices: the manufacturers usually aim to cover the entire market with the same product version and thus need to generalise and abstract information. This is expressed in an engineering language. Pump controllers, for instance, show Hz in the display, even though water plant operators would need to see the pump speed.

At wastewater treatment plants heavy rainfall poses a serious problem to the process, due to overflow in the basins and poor de-segmentation of the sludge. In the water vision project we used the rainwater de-routing problem as a starting point for exploring the design of future intelligent water components.

The subsystem involves flow meters, sensors, pump controllers and controlled valves. With modern sensors it is possible to check the water quality on-line in the pump stations and thus detect rainwater before it reaches the wastewater treatment plant. Pending municipal authorisation, it should be possible to de-route the water in pump stations as soon as it exceeds the quality that the plant can produce.

There were ten of us in the project team: user-centred design specialists, developers from business units, management trainees, and university students. In total the project took ten months. We gathered to explore the idea of whether components could negotiate. Could industrial components like those of Danfoss manage the task of rainwater de-routing all by themselves, if they were communicating on the same net?

The traditional approach to pump station design is to program a programmable logic controller (PLC) to work as a central controller. This means that the PLC contains all the application knowledge of this particular pump station design. The system must be designed and completed by a specific time. However, the reality of most plants is that their instrumentation and control keeps changing with new technological opportunities, tougher regulations, *etc.* With new kinds of technical opportunities it may be possible to realise a control system that is continually updated, when a new component is added to the system. If this system were created, with each component incorporating application knowledge sufficient to negotiate the task in a net of distributed intelligence, what would the operator's role be?

To learn about the pump station, we used field study outcomes and built a copy of a pump station out of cardboard and other materials. This understanding served as the basis for the ideas that we further explored through the means of an enacted scenario. A basement room served as the physical stage where the scenario was acted. Through the scenario-acting we learned how important it is to build application specific knowledge into sensors. With such sensors, new knowledge could be added to the plant. New sensors could measure a wider spectrum of nutrients and their concentrations to improve the process.

The pump station video was never produced to be shown outside the team. The camera simply acted as a means to discipline the acting, to be serious about it, where the main purpose was really to learn about pump stations, autonomous components, and networks. ∎

"I would use it this way"

Acting with props is a method that utilises a rough representation of ideas as a starting point for exploring how an idea might work in a concrete use situation. The participants create one or several scenarios that revolve around the use of the prop. *Prop* is a term borrowed from theatre. It refers to an artefact created to give the impression of it being something else: a suggestive artefact. For example, a box on a bookshelf may signify a television. In a play it might be a dagger, a letter, a glass, a coat. Props are crucial tools for the actors to develop the play and create suspense. In design, props are mock-ups or prototypes of product ideas, created of materials such as cardboard, foam, stickers, clay, construction sets, *etc.*

The prop focuses the design activity by proposing a concrete shape with which to work. Svanæs and Seland (2004) discovered that design ideation with lo-fi prototypes, *i.e.* mock-ups, requires some constraints. The ideation is otherwise likely to produce non-relevant ideas, and the process may appear difficult and demanding. A concrete environment and a realistic situation possess such constraints within which the prop can trigger questions to be answered. Donald Schön (1983) speaks of "backtalk", *i.e.* that the surroundings "talk back" when faced with a new design creation.

When using this method with users, the shape of the prop is often intentionally left very rough and open for two reasons. First, the rough and open shape allows greater freedom for imagination. Different people may think about the shape differently, and this contrast of interpretations may trigger new ideas. The rough shape invites users to modify and complete the shape in a direction that the user finds exciting. Second, when the shape appears rough, the prototype is also seen as a rough idea. If a rough idea is presented with a polished appearance, it is likely to invite premature evaluation of details like the size and placement of buttons when the design concern is actually on product functions.

Staging the event

Design the props. Prepare a set of design props appropriate to the stage of the project. Often preparing more than one design alternative can help participants to adopt a prop, if they are encouraged to choose the one they prefer. This in turn encourages discussion on why. Alternatives may also be introduced later, to develop the scenarios: how would the same situation work with this prop? One possibility is to create a set of make tools (Sanders and Dandavate, 1999), as in the "Ageing future"

case in Chapter 2. Such Velcro models are configurable and reusable
for different projects.

Provide a background. Materials from background studies may be presented to help ground ideation on real situations.

Introduce the props. Find a way to familiarise participants with the design props. There are many more engaging ways than a long verbal presentation. For example, in one of the workshops organised for authoring this book, the participants received a dummy book cover and were asked in small groups to explain why they had bought this book. This encouraged them to talk about themselves and their interests.

Construct a situation that can frame the acting. This may happen before the event or as part of the programme, based on participants' suggestions. Make sure there is a challenging goal to work towards. "Goal" can here be understood on two levels: the *overall aim* of the activity, and the *user's goal* in a specific exploration. The overall aim is, for example, to search for alternative physical shapes for a mobile controlling tool for a janitor, and the user's goal could be, for example, to adjust the ventilation in the building with the imagined tool.

Explain the rules. If the session is improvised in the manner of a game, rules are needed to outline what is allowed, needed, and desired, and what is prohibited.

Create the environment. Prepare or decide on an environment that may inspire acting. In a lab, a cardboard stage, for instance, helps to ensure a playful atmosphere. A session in a use context sets the expectation that this is serious work.

Assign participants roles. Who shall be the bartender? Who shall be the client? This part is often loaded with enthusiasm and engagement, and extra props like a hat, a lab coat, or a pair of gloves help people to take the part. Often it makes sense to let participants act themselves. In this way they can bring their "natural" ways to initiate activities and respond to situations to the interaction.

Encourage ideating. This may include practical tips, such as encouraging building on each other's ideas, constantly trying things differently, making as many new attempts in a short time, *etc.*

Use the video camera to focus the activity on when to act (camera on), and when to discuss (camera off). This is like the one role of the clapboard in traditional moviemaking – to create concentration.

Reflect. Reserve time to view results and reflect on the scenarios created.

In the making of the scenarios the whole reality is fluid. A design team may replace a world with a new one in the matter of an instant. The props help the acting focus on concrete proposals about what reality would be like if it proceeded within the suggested lines. When employing improvisation and scenario acting, the design team needs to have their heads in the clouds and feet on the ground. It needs to embark on a road towards good ideas with the confidence that the process will develop such. This, of course, cannot be foreseen, as it results from collaborative construction and does not yet exist. As Keith Johnstone writes:

> *...the ideas that emerge in the spontaneous improvisation may be irrational until the next ideas render them sensible.*

Ideas are always unique to the moment of time, to the project and to the people present. ■

Ethnography of the future

Iacucci, Kuutti and Ranta (2000) suggest that the situated participatory enactment of scenarios may be understood as conducting "ethnography of the future". This underlines their understanding that "mobility issues are best studied when on the move, and that personal matters are best observed in personal situations." (Iacucci *et al.*, 2000, p. 200).

Would it make any sense to think of video scenarios as ethnography of the future? Ethnography is the traditional method of social anthropology to study and describe a human community. The word "ethnography" also refers to the written "thick description", or theory, that explains the studied culture. If we think of video scenarios as ethnography, we need to understand these (re)presentations somewhat differently from traditional ethnographies. First, the result is authored by designers and users rather than cultural anthropologists. Second, the presentation format and relationship to theory-building differ greatly. Blomberg *et al.* (1993, p. 143) write that:

> *Understanding and insights derived from the study would not necessarily be represented in a written report, but instead would be reflected in a codesigned artifact.*

Third, the focus of these representations is to delineate a desirable change, or a desired state of affairs, rather than to explain a community of practice.

▶ Case story: The social microwave
Kyle Kilbourn, University of Southern Denmark

Video
scenario
**The social
microwave**
1'23"

"We don't have a name yet!" I say, panicking at the thought of giving a presentation of our nameless design concept in less than ten minutes. While I have been stressed about piecing together the interactive poster, Jan, Sarah and Shirley have poured many hours into sculpting the concept from thoughts in our heads into a Wizard-of-Oz prototype of a microwave of the future. Knowing Sarah wants to emphasise the effect of lifting the lid and presenting the food, I suggest "Tada." It is a common American expression used when you want to be melodramatic in presenting something special. No serious objections from the other team members, which is good because it is show time.

With Shirley manning the computer to control when the animation changes colours and I struggling to hold the mirror we "borrowed" from the men's bathroom to reflect the projection onto the surface of the table, Jan and Sarah are left to act the part of the couple cooking with the futuris-

For example, the "Nokia: DrWhatsOn Concept" shows life with a mobile phone that could automatically switch to mute mode in a library. Fourth, these presentations are "thick" only to the extent that is necessary to promote the arguments for the value of the product. In the "DrWhatsOn" case the video scenario does not explain why the person goes into a library, or what the meaning of the library is for the person or to his community. It only illustrates how the new product keeps up with the pace of events. Hence, if video scenarios are understood as "ethnography of the future", we are faced with a completely different practice from traditional ethnography.

The "ethnography of the future" as outlined above is grounded on making people act. Film-makers have long known how well people are able to

---

tic microwave. They play the role like the perfect couple, even though they have only known each other for the last two weeks.

Sarah: *So I just put it on the tray, and then put it in our microwave. Top
         on. So, these are only vegetables, right? What heat should they have
         then? I think orange, maybe?*
Jan: *Women today don't even know how to cook.*
Sarah: *Okay, blue is like, cold. Yellow is not enough...*
Jan: *It should be more like... [Jan takes control and twists the handles.]*
Sarah: *No! I don't think so. You'll destroy the vitamins. I think we should
         set it to orange.*
Jan: *It should be short and hot.*
Sarah: *But still it is vegetables. I think we should stick to orange, right?*

Dialogue like this makes the audience chuckle at the situation we present, even anticipating the next bit of comic relief involving the social microwave. Jan and Sarah play to the camera, leaving behind their original design intentions of being elegant and presenting the food in a sophisticated fashion. Interacting with the Tada microwave is a reward in itself.

Inspiration from the past and the present to create the future. Our microwave came from two intensive weeks in March 2004, while investigating the interaction styles of kitchen products throughout history. In the first few days, about re-enact their lives.[†] Designers have also successfully employed real users as actors in a diverse variety of video scenarios. For example, Sperschneider and Bagger (2000) present a method where real users re-enact past situations. When people act out their past experiences, they will bring their usual ways of coping into the situations. This is evident, for example, in the "Lapland hiker" case story in Chapter 5. In this scenario the novice worker is acting as himself and improvising according to a roughly planned plot. The improvisation by four different workers during the study revealed details of both the working culture and the workers. Enacting future scenes poses a dilemma in studying people's practices, since the future situation is new also to them. What can a design team expect to find with such an artificial setup?

† Flaherty's Nanook of the North (1922) is a great example of a classic documentary film where this method is utilised.

20 Master's students in IT Product Design at the University of Southern Denmark sought inspiration from museums in Bjerringbro and Horsens, Denmark. Armed with photos and video clips of products from the past in action (luckily, they were "hands-on" museums), a group of students painstakingly split what we saw into several style periods. These periods described in some detail the students' view of the major influences on society, technology, hand-actions and space within the realm of the kitchen.

The four of us (Shirley, Jan, Sarah and myself) needed to design a microwave concept inspired by the final style period from the 1990's and beyond, called "induction cooker playboy." Impressive, intelligent, cooking without touching, and subtle were a few of many keywords meant to guide us in the design process.

Sharing a common vision. A challenge in the project was reconciling our own understanding of what we were designing. This may have been a combination of our various backgrounds and cultural differences. Shirley, a Chinese computer science student, thought of the microwave in terms of a thermal imager because of the way it was to project the temperature of the food onto the lid. Jan, the Dutch designer, and Sarah, the German with a background in cultures and languages, were set on thinking of it as a silver platter, giving it a touch of class. While I, the American that had studied biology, wanted to emphasise its interaction as compared to a stove.

At a mid-critique a week-and-a-half into the project all we had was a large plastic bowl to show for it. Quickly trying to summarize our thoughts so far, I

New media technologies will change the communicative practices of people. Social organisation in situated interaction is accomplished through means such as bodily expressions, responding to responses, orienting towards objects of interest, sensing, talking, moving in space, and using physical things. Any of these aspects may be changed as a result of introducing the new design into the setting. There are simply so many changes triggered by the introduction of new design that a design team has little chance to foresee the true influence and value of their proposal. When design moves towards the production of the intended change, designers increasingly need to gain awareness of the impact of their design in order to ensure the delivery of *good* solutions. Only such awareness

wrote down a few words we all agreed on: "advanced dinner, experienced cook, on/off with lid, turn pot to control heat, microwave like stove–not oven". This captured our thoughts about the microwave, but did not accurately symbolise what would become known as the social microwave.

Constructing meaning through acting. As we were rushed to physically design and construct the Tada microwave, we forgot to leave ourselves time to come up with a brilliant script for the final presentation of the concept. In the last ten minutes before the presentation, we had an impromptu discussion that led to the name and an assigning of roles in the acting out of the scenario.

It was during acting out in front of the video camera that Jan and Sarah let loose and did what they felt came natural when interacting with the Tada microwave. The earlier notions of elegant and refined dining just did not feel right.

The discussion immediately after the scenario allowed the audience to vocalise their understanding while at the same time questioning us on our own interpretation of the design. Several themes popped up: the dilemma between the time needed to prepare food and the socialisation that occurs, the quickening pace of modern life and how to get a social feeling in a shorter amount of time, and the impact of globalisation and internationalisation on the kitchen. While we may have shaped the physical appearance and interaction of the microwave, it was our acting with the microwave that ultimately shaped the shared meaning with our audience. ∎

enables them to judge whether the perceived changes point in a desirable direction.

Jeanette Blomberg and her colleagues (1993) argue that it is important to link the ethnographic study of current practices with the involvement of the users in developing the new designs. Through such cooperation it is possible to gain new understanding of the evolving practice. The key question here is: "How will the *new* digital objects mediate the multitude of discourses that people engage in with their environments and with each other?" On these grounds it is reasonable to assume that actions should be taken to understand the true changes that designs impinge on people's lives.

A chicken–egg dilemma

Knowing what kind of design would be good for users is a chicken–egg kind of dilemma. In order to see how the new practice becomes influenced by the planned designs, these need to be placed into the use context and how the practice is changed observed. However, in order to conceive a new design – one that can be expected to function well for the users – designers need to understand the changes it imposes on the practice. What possibilities are there for designers with video equipment to tackle the challenge? The key issue is to understand that, rather than prototyping the technical functionality of a product in the early phases of the development, designers need to "prototype social action" (Kurvinen, 2007).

In the early phases ideas are usually iterated (created, tested, evaluated, and modified) in a rapid cycle. With video this means the hasty staging of scenes, playing ideas out, and seeing how they function. The staging involves materials such as cardboard mock-ups, papers, pens, transparencies, existing devices, *etc.* The emphasis on how the stage becomes set varies across methods. For example, situated and participatory enactment of scenarios "spes" (Iacucci et al, 2000) emphasises the value of everyday life contexts as the "stage" for the improvised scenarios, whereas Binder's (1999) "improvised video scenarios" focus on the utilisation of props without explicitly framing the goals of the design event for the users that improvise. On the contrary, Ehn and Sjögren (1991) promote the value of setting the props and building a physical setup in order to establish a common language that both users and designers understand and are able to use. Other methods include, for example, design games (see *e.g.* Schuler and Namioka, 1993, Greenbaum and Kyng, 1991).

In theatre acting takes place on a stage. This is the physical arena with dedicated materials that outline the physical borders of the play. However, the physical setup forms only part of the staging. The other part consists of the staging of thinking. The mental setting is perhaps even more important for grounding the collaborative acts of imagination. As stated in the previous section, the staging needs constraints, or "givens". The givens help people gain a similar orientation to understanding the situation and seeing what ought to be done in it.

The above methods are mainly ways to set up an event that facilitates the enactment and exploration of the possible new practices, *i.e.* causing the change. The other side of the coin is to study how the features of the designs influence the practices and why, *i.e.* investigating what happened. This is where the theorising about the issues, reasons and effects, or the "ethnography of the future", begins. Here the scenarios gain a stronger role as the catalyser to see oneself act – to borrow Boal's ideas from theatre.

Theatre of the Oppressed

Augusto Boal was annoyed by the way theatre was employed in society. Boal was especially struck by the elitist character of theatre. The ordinary people from São Paulo, where Boal lived during the 1950s, were not theatre-goers. Theatre was conceived by Boal to be an important means of social and cultural influence, and it should therefore be available to everyone. When he developed the Theatre of the Oppressed during the 1950s and 1960s he was devoted to the idea that theatre would turn into more *dialogical* practice, in contrast to its long history of monologue (Boal, 2000). Theatre plays used to be acted out by professional actors to the audience that perceived it rather passively. Boal held the attitude that dialogue was the healthy dynamic between all humans, and that all human beings are capable of, and desire, engaging in dialogue. When the dialogue turned into monologue, oppression would ensue.

Boal's approach focussed on democratising theatre and empowering people to affect social change (Boal, 1998). He thought that theatre must evolve into a tool that enables transforming monologue into dialogue. He wanted to *create the future* with people, not to wait for it passively. Boal developed numerous approaches to developing dialogue in theatre. He created workshops that aimed to foster critical thinking, interaction, action and fun. Boal's methods, such as Image Theatre, Invisible Theatre (Boal, 1992), and Rainbow of Desire (Boal, 1995) aimed to bring the audience into an interactive relationship with the performed event.

Boal's workshops typically comprise three kinds of activities: first, providing background information about the Theatre of the Oppressed, second,

using games to sensitise people to listen to what they are hearing, feel what
they are touching, and see what they are looking at, and third, applying
structured exercises utilising some of the methods, such as Image Theatre,
Forum Theatre, or Rainbow of Desires.

Image Theatre works to depict concrete images of how people conceive
their reality. The method begins from individual participants' (or the "spect-
actors", as Boal calls them) images. They present these images by "sculpting"
a static posture with their bodies. The image may represent any theme, for
example, "the family". At first, one of the participants works as the "sculp-
tor" to create the picture. Then, if the other participants do not agree with
the image, they are asked to refine it. The process continues until the group
has settled a consensus about the image. This first image presents the current reality, or the *real image* – as Boal calls it.

Sometimes, especially when people do not know each other, Image Theatre can be utilised to embolden the "spect-actors" before starting Forum Theatre. Boal (1992) thinks of Forum Theatre as a "fight" or a "game". Forum
Theatre aims at provoking responses from the "spect-actors" with a play that
displays an apparent conflict. When the play is presented a second time, the
"spect-actors" (*i.e.* the audience) are asked to stop the play when they perceive
a mistake is taking place in the play. Then the one who asked to stop the play
will take the position of the protagonist and start to direct the group of actors,
who have frozen in their positions. The new protagonist attempts to correct
the situation while the actors keep fighting for their previous ways of going
about things. In this way Forum Theatre helps to make issues as concrete as
possible for everybody to discuss, elaborate and try out variations.

A fundamental aspect in Boal's theatrical methods, as well as those presented by Johnstone, is the breaking of the traditional frames for thinking,
the habitual ways of acting, and to heighten the senses – escaping deeply
rooted routines. This moves the thinking towards perceiving reality in a new
way; we could say that here the perception transforms the reality into *designer
clay* that can be moulded into new forms.

● Method: On-site Scenarios

On-site scenarios feature the usual environment of the users as the stage
for the acting. The environment helps bring the knowledge of the users to
the scenarios in a rich form. For example, the availability of all the tools for
easy reference allows for fluent consideration of the ideas in relation to the

*"Now that
phone
rings..."*

contingencies that arise out of the everyday spaces. The normal environment also – in a sense – forms a benchmark against which the new ideas about an improved practice become almost automatically compared. These issues facilitate the development of ideas that have true potential to improve the practice of users.

Being at the users' site is likely to affect how the users feel about the scenario-making situation. When the scenarios are acted out in the users' environment, the designers will be the visitors and the users the hosts. This often helps users feel more comfortable and encourages them to maintain control over the collaborative exploration of ideas. Being at the users' site also affects how designers look at the activities. People's interactions are highly attuned to the details of their usual environments, whereby the improvised exploration of potential situations reveals much about what the users bring to future interaction with novel solutions.

The real environment fosters the creation of scenarios that have a sense of realism built in. If the ideation becomes grounded on a review of earlier experiences of relevant situations, as in the case "Phoning a deaf person", the scenarios also gain higher credibility with realistic detail. Moreover, when the users are involved in the authoring of the stories, the story becomes immediately verifiable in relation to the users' reality. The story may evolve very quickly towards a relevant innovation compared to a situation where a designer, foreign to the users' reality, would imagine it. This is why Sperschneider and Bagger (2000) also promote the idea of design-in-context.

Issues to help focus on-site scenarios

Acquire permission. The working sites of users may have restrictions on videotaping. Hence, on many occasions the design team needs to negotiate where and how the video scenarios may be created.

Inform others. If the shooting is planned in a place with people who are not participating in the scenario-building, they need the opportunity to escape the video camera. This may require reconsidering the location for videotaping.

Prepare props. As in "acting with props" on-site scenarios may utilise props to help imagine how the ideas could be utilised at the site. These may require some preparation to align them with the project aims.

Create a situation. Acting is much easier when the actors can imagine a concrete situation with which to work. This may require thinking

about when the scene happens, what has just happened, and what the
user might have to achieve in the situation.

Several of the case stories in this book were captured in the users' real environment. For example, the mask scenario in the "Intelligent pump station" case was later re-captured at the real wastewater plant. While the environment was utilised in the scenario somewhat differently than would be expected by the method description, the environment provided a familiar place for all the operators and engineers who participated in the acting. Moreover, it fostered a more accurate feeling of designing intelligent devices for a real wastewater plant, as the acting was carried out in the middle of the pipes and valves closely related to the work.

The case "Lapland hiker" in Chapter 5 also displays a variation of the method. The acting was conducted in the real facilities, but the shooting was located in a different room to avoid disturbing the real telephone consulting activities taking place in the real phone service room. The acting was assisted with a rough plot that was planned before starting the improvised acting.

Encouraging the users to improvise and develop the ideas is a means to focus on those ideas the users feel are important and should therefore be most relevant. When such improvisation is organised at the users' site, the environment facilitates comfort, inspires and guides the design. It also helps to construct visual material that may provide designers with visible arguments for design in the later phases. ■

Directing the future

A controlled process, like the one usually adopted by professional movie directors, also has value for design. Manuscripts, in their various formats, are the backbone to a systematic way of creating videos. The writing down and the formalising of large and detailed ideas makes the planning of the whole a lot easier. Alan Rosenthal (1996), a film-maker and theorist, parallels the manuscript with the architect's plan, however, with the precaution that the manuscript may still go through substantial changes on the desk of the editor. We shall learn a bit more about the reasons for this later in this section.

Why should designers invest their resources in creating such detailed video scenarios? Scripted video scenarios depict details of well-chosen moments in future. Entering an imagined future is rather easy, as we learned in the previous section. However, entering a *potential* future is somewhat more challenging. Moreover, familiarising with a future well enough does require a bit of work. This is where the controlled process of authoring the scenarios becomes valuable. However, what does it mean to familiarise with the future "well enough"?

User-centred design promotes the understanding of the use context. Designs are perceived as good or bad in relation to how well they fit to the users, their aspirations and their ways of going about things. Knowing a future "well enough" thus presumes developing an understanding about how the proposed designs function in the active world of the users. A video scenario is perfect for this.

How, then, do designers identify a potential future worth making a scripted video scenario? Often with design scenarios the story develops in the course of the collaborative exploration of the specific area of designing. The process of developing the story for video scenarios is usually a collaborative study of the design opportunities, and may involve a study of the users' current practices. The co-building of the story as a manuscript or a storyboard provides users, also those not willing to act, with a means to participate in scenario building.

Books on film-making, scriptwriting and directing usually outline a three-phase structure for the work: pre-production, production, and post-production. Roughly speaking, pre-production contains all the activities that prepare a film crew for the shooting of the acts, and post-production includes the work when all the material is "in the can". This is an extremely rough overview of the whole. These activities will be briefly described here, with a focus on activities that are seen as useful for designers. For the reader interested in more details about the whole process, we recommend several excellent books such as Michael Rabiger's "Directing the Documentary" (1987) and Alan Rosenthal's "Writing, Directing, and Producing Documentary Films and Videos" (1996).

Rosenthal begins his book about film-making with a scene where he is discussing with his colleague a possible documentary movie they may create next. Such a discussion, like the one active during the product concept search, may lead to activities of making a documentary, or making a product – an endeavour that may last from months to years. Rosenthal emphasises that there is one vital question that needs to be answered before committing to anything: "Why do we really want to make this film?" (Rosenthal, 1996, p. 7)

If this question has a decent answer (perhaps even as simple as, "we would like to try what an exciting new video might provide us"), the next step is to consider the feasibility of the idea. "What scale would be appropriate with the resources we have?" Depending on the scale of the project, the process will comprise different kinds of activities. In the extreme case, like "Starfire" described by Bruce Tognazzini, or the case of "It's UI Love" (in Chapter 5), the process is basically similar to creating a real movie, as outlined by Rosenthal (1996, p. 12): (1) script development, (2) pre-production, (3) filming, (4) editing, and (5) final lab work.

Script development begins by developing the idea for the scenario. The case story "Context aware mobiles" provides a nice example of the initial outline of the idea, the synopsis. The first phase may also include discussions with sponsors and funding agencies, preliminary research, writing a proposal, discussing, agreeing on a budget, research, writing the shooting script, and accepting the script (probably with a number of modifications). According to Rosenthal (1996, p. 10) the manuscript plays a number of roles in the process: *it is an organising and structural tool*, which serves as a reference and guide, and *it communicates the idea* of the "film" to everyone involved in a clear, simple and imaginative way. For the camera person it conveys the mood, action and issues related to capturing, and it helps the director to define the approach, the progress, the inherent logic and continuity. *It answers the questions of the film crew*. These include issues such as appropriate budget, locations, lighting, special effects, the use of archives, and special equipment needs. The script also *guides the work of the editor*. However, the editor may create an "editing script", which may differ quite radically from the initial script.

How does the idea for the story develop? For example, in "Helping the hiker" the idea was grounded in a contextual study at the workplace of the phone service attendants. The situation for the scenarios was chosen based on several influences: first, the aim of the project to develop new concepts for knowledge management which outlined the kinds of situations that would be explored; second, the results of the contextual study; third, the ideas that the workers expressed when they were told these results. The idea of the service was designed along with the story. The project began with two scenarios about the current situation, which provided grounds for evaluating the development potential that was concretised in the later two scenarios.

Moviemaking is usually taken as a fun and motivating activity by the users, designers and other participants in the video-making events. The ideation sessions where future situations are played out form engaging events, which people may remember for the rest of their lives. The enthusiasm in co-authoring a script can also surprise the designers, as they see how much creative energy there is in "everyday people".

● Method: Scenario Scripting

*"Then he
will open
the door
and say…"*

Fundamental to the use of scenarios is the quality of bringing design ideas into the context of potential situations in action. This enables a study of how the ideas influence the context and what the design ideas need to respect. The activity of negotiating a story helps uncover the issues relevant in the interplay of changes between influenced human practices and between the proposed design ideas. Scenario scripting is the activity of systematic planning and study of these changes with the tools of playwrights.

What if designers were to act as professional playwrights to make the story? This has actually been attempted on several occasions both in academic and industrial contexts.† Professional scriptwriters excel at crafting exciting stories that convey definite messages. They have the understanding of how to capture people's attention with the sensual exposition of surprising, thrilling, and telling details that move the story forward. They are able to populate the story with emblematic details of everyday life and, as professional storytellers, can produce credible stories in a rather brief time.

This was attempted, *e.g.* at the University of Art and Design Helsinki with students of film scriptwriting in 2002

These skills make playwrights an excellent aid in the exploration and description of the potential relationship between people and designs. However, the risk with professional scriptwriters is in moving the focus from conveying the value of a design to exploring the suspense and conflict in the story – the misery or glory of the life of the protagonist. Design scenarios, to the contrary, should focus on how the new products bring value into people's lives.

This is where the collaborative authoring of the stories becomes helpful. The availability of the skill and experience of different people fosters a deeper level of reflection on the relevant issues relating to the quality of the product. Since design scenarios are most often created without the assistance of professional scriptwriters, the following list of guidelines aims to help a design team craft effective scripts for video scenarios:

Provide a context at the start. The context helps the audience to understand what the story will convey. It outlines the initial situation, the environment, the people, and their aims.

Flesh out concrete details. The facts from user studies and personal experiences help build credible stories. These details foster user empathy, support making engaging and interesting stories, and may even inspire new ideas.

▶ Case story: Smart packages

Kirsi Kauppinen & Daranee Lehtonen, University of Lapland Rovaniemi

Susanna is shopping for groceries in a large supermarket with a wide product selection for vegetarians. She picks up a can of soup displaying the symbol of global ecology that stands for sustainable development. The 3D symbol rotates as she turns the can. As she passes the coffee shelves something captures her attention. She steps over to take a closer look. A set of coffee packages compose a big graphic surface. A new chocolate flavoured coffee has been launched. Animated steam smoothly dissolves on the package surface. Susanna's face lights up. I'm going to treat myself with a warm cup of chocolate coffee!

The Printo project developed methods for mass-producing optic, electronic and optoelectronic components that can be cost-efficiently integrated into packages and printed commodities. The project was part of ELMO – the Electronics Miniaturisation programme of the Finnish Technology Agency, TEKES. Printo started in April 2002 and focussed on developing concepts for smart packages. Multi-layered recyclable cardboard would enable various new uses for the surface of the products. The new technology had the

Video
scenarios
**Printo
scenarios**
1'12"

Show the value – do not tell it. Video is a medium that is best at showing how things happen. This quality makes it perfect for illustrating how the product ideas would function in the future practice of users. The concrete image also conveys things without the need to explicitly tell everything in words. For example, if something makes someone sad, the image of the sad-looking person is enough to make the point. Similarly, if the message is to illustrate how practical an idea is, it is better to show it in action than to explain the value in words. Showing instead of telling also leaves the delight of conceiving the ideas to the audience.

For example, the scenario could start by showing a situation where Diana, a 35-year-old accountant, looks out the window and sees people waiting for a

capacity to increase the information content, enhance the appearance, and add interactivity with users and with the other intelligent appliances near the product. Our task at the University of Lapland was the conceptual design and visualisation of the new technology's commercial applications. We generated smart packaging concepts in close cooperation with the technical research units in the project. Our role was to improve the design communication and bring user-centred design into the project.

Before we started to shoot the video in April 2003 we wrote a manuscript for the 16 scenarios for three time ranges: the year 2004, the years 2005 to 2009, and 2010 and beyond. The scenarios were based on product concepts that were developed during the first phases of the project. We thus created the scenario situations based on the product concepts; we had, however, conducted the user study earlier, which provided the appropriate background for constructing the situations. The stories aimed to illustrate how the new kinds of products would provide users with new kinds of values. We wrote the original manuscript in plain text, but before the shooting we also crafted a storyboard to help production. The scripting and storyboarding took in total two weeks.

The shooting was quick compared to the pre-production work. A day was spent in planning and scouting the locations. We spent two days preparing the mock-ups for the scenes, and three days for the shooting. We recruited colleagues, friends, relatives and students to be the actors in the scenarios. The scenes were explained to the actors on location. With the storyboard we illustrated how the situation unfolded, ran the camera, and captured the action. The tram on a rainy platform. She hangs her coat that she was about to put on back on the peg and takes a look at the traffic display on her desk. It shows a tram in a traffic jam two kilometres away. She sits down, leans back in her chair, and makes a call to her friend as there is a free moment for a small chat.

What is the *value* of the product (traffic display) here? How is it *shown*? The product enables the woman (1) to avoid getting wet, and (2) to utilise her waiting time by doing something else. The first part of the value is expressed through depicting the unpleasant consequences, *i.e.* the people standing in the rain, which would result if the new design had not helped to avoid it. The second part is visible in the manoeuvre of leaning back in her chair and chatting on the phone.

scenarios did not include any dialogue. The acting consisted of moving, looking, orienting, and picking up the objects in question.

To minimise disturbance at the shops we planned the shooting for midday when there are fewer customers. We also allocated roles to make the session efficient: one of us operated the camera while the other was directing the actors and placing the mock-ups. Because of the haste and continuous attention on the periphery, some scenes failed to provide us with proper material. We decided to cut the troubling parts. Filming in homes was much easier, because we could retake the scenes as many times as we needed.

The frustrating post-production. The post-production was tedious as we did not have any previous experience in video production. We needed to learn the software for editing and for animation, and we had to study how video conveys the story effectively. We also had to search for copyright-free sound effects. The looming deadline of the concept validation with the users made us feel quite uncomfortable. Finally, after two hectic weeks of editing and animating, the scenarios were ready.

Provoking global debate. Video scenarios were used later in concept validation to gain understanding of how users feel and think about the developed scenarios. During this phase, we used 7 small focus groups to validate the concepts with altogether 21 heterogeneous users in Finland, Germany and Thailand. Packages are global, and it was found crucial to explore the cultural differences

One trick that designers may utilise to help convey the value of a product is to create contrasting scenarios. This means that first the current state of affairs is explained in a scenario. It helps to understand what the practice currently entails, and makes it possible to evaluate how it becomes improved as the result of the introduction of the new design into that practice.

The two Nokia case stories were built around the idea of context awareness, and the stories developed in a debate about what the situations would actually be. The process of collaborative authoring of the video scenario manuscript helped to crystallise the core value and character of a new product idea. For example, the case story about Nokia's DrWhatsOn 11 in Chapter 5 shows how the design team struggled to negotiate the core value related to the subject. The aim was to obtain diverse opinions related to the different use contexts of the concepts directly from the users. Video scenarios were useful in provoking debates during the user sessions.

The right medium. We found video to have quite a high value in the work with users. It helped to overcome the language barriers – also because we avoided the use of dialogue in the story. We considered the use of computer animations as well, but video was far better in presenting the context in a credible fashion. Video enabled us to visualise and concretise the new, future-oriented technology in quite a realistic manner. It displayed how the future concepts work and for whom they are meant. Moreover, video illustrated in a condensed and concrete format the abstract theme of "smart packaging" and its value for people.

Despite the roughness of the final video, something that needs to be considered in future projects, it helped us achieve what we wanted. The concepts were quite open – only the overall design was presented as the focus was placed on the context. This turned out to be an advantage, since it gave the participants the feeling of early and conceptual ideas. The feedback addressed relevant issues in the work.

The analysis of the results from the user evaluation of the concept ideas with the scenarios guided the activities during the third year of the project. The project then focused on developing illustrative demonstrations and prototypes of the ideas. ∎

of the new design. By being forced to convey the message in a simple and condensed format the story helped the team to also develop the product concept further.

In the "Lapland hiker" case the initial acting of two roughly sketched and half-improvised scenarios about the current practice provided the design team with grounds to plan the details of the future scenarios. The collaborative construction of the story was especially fruitful in promoting a consciousness of the relevant issues when designing useful new knowledge management practices in the telephone banking service.

Scriptwriting encourages discussing alternatives ways the activities might unfold. This promotes the development of a multi-faceted understanding of the design situation in question. This is quite different from scenario improvisation, where the story unfolds chronologically from beginning to end. Moreover, when the designing is in the form of manuscripts, it is still very easy and quick to change the ideas. In this phase reality is still fluid, and the cost of large changes does not radically increase until the later phases. ∎

Co-creating

Concrete images of possible futures enable the making of judgements about what would be preferable. Video scenarios are concrete illustrations of what reality would be like if it were resolved in "this way". When ideas are expressed in the language of human practices they make it much easier for the designers, managers and users to evaluate that this is indeed desired.

The collaborative authoring of scenarios plays several roles in the user-centred process of designing. Firstly, improvised acting is great fun. Having fun together is an excellent means to build bonds between people. Nice memories from co-improvisation events create a pleasant background for future collaboration. Secondly, video is perfect for displaying realistic stories about how products fit into human practices. Video scenarios place new designs into the practice of people and enable the study of their impact. Hence video renders the interactive relationship between people and products within realistic environments visible and mouldable. effecting this way, video transforms human practices into a kind of clay that can be purposefully designed. When combined with a properly set stage that brings findings from the background studies into the design events, the making of scenarios may establish an effective co-creation of futures and a fruitful dialogue between ideas of change and experiences of the past.

[The upper portion of the page contains a photograph of a mobile phone being held, with the text "8244" and a name label at top, a message reading "Out for a coffee break, back soon -Mike" overlaid, and the caption "locating Mr. Smith" at the bottom.]

Video
scenario
DrWhatsOn I
7'47"

▶ Case story: Context aware mobiles

Urpo Tuomela, Nokia Corporation

It is a dark and rainy day in mid-March, and we have our first shooting day. We had planned for three episodes to be shot during this first day. It could have been a bit better weather for outdoor shots, but because of the tight schedule we have no alternative. The first shot is about the ticket payment when entering a bus. We use two cameras to cut down on the number of takes. We also streamline the audio recording by using a narrator instead of real dialogue.

Mika takes the first camera into the bus and negotiates with the driver about the needed shooting activities. The bus is a regular commuter bus at the Technology Village going towards Oulu city centre. The driver lets us take all the needed shots during the trip. I am waiting for the bus to arrive at the bus stop with our lead male actor (later referred to as "the dude") and another camera. The rain is not too bad, so I am able to use the camera outside without extra covers.

The DrWhatsOn project started in January 2000 and ended in November the same year. It originated in the technical research of context awareness, which we had started in 1997 while participating in an EU project called TEA (= technology for enabling awareness). In the TEA project we studied technologies and methods to automatically recognise the varying contexts of people's use of mobile devices, and the possibilities of these devices to adapt to these contexts. DrWhatsOn aimed to take these ideas further to explore how people can benefit using context aware technologies and applications. We also concentrated on issues related to the user interface. The focus of the project was to create a concept for a context aware mobile device for people in office environments. The video also aimed to summarise the "state of the art" knowledge of context awareness technology. The video was to explain the technology in an understandable way and illustrate how the technology works with applications that make sense in everyday life.

We started the authoring of the video by outlining the synopsis:

"A dude walks around with his mobile phone and does a lot of different, ordinary-life things. The camera follows him, and his thoughts are spoken aloud in the soundtrack. In the end the dude leaves his mobile phone behind, on a desk perhaps; the camera zooms onto the phone, and the viewer realises that it was the phone that had been thinking all the time!"

I wrote the script together with my colleague, Petri, who kept the inspiration going and took care that all original ideas were taken into account. To improve the script we circulated it within our project team. We then formed a video team for the first video, and went on to planning a schedule for the video making. The final script consisted of seven episodes. We planned the shootings in March, editing in April and the premiere in May. The schedule was quite tight, and we needed to work fast. For example, a draft storyboard was created but was never finalised due to the tight schedule.

After a short and intensive search we decided to shoot three episodes at the University of Oulu, one episode outdoors and in a local bus, two episodes in the Nokia premises, and one in my own apartment. Considering the tight schedule and limited resources we decided to proceed in an economical way – investing minimal resources. For instance, we decided to shoot everything in natural light.

The shooting resulted in some 50 minutes of video footage. We watched it with Mika to identify the best shots. Then we began to work with the narration, which was initially planned during the scripting. After modifying the texts to fit to the result, I asked a colleague from the us to help us with the narration. He watched the video a few times and read the narrative texts to become acquainted with them. His elegant voice gave a completely new depth to the DrWhatsOn video.

When the video was further supplemented with the sound effects and music created by Mika and Schubert, the video gained yet another layer of professionalism. Together these sound specialists authored the audio details to serve the needs for display on computers and in auditoriums.

Good planning enabled straightforward shooting. However, despite the good plans, we had to improvise quite a bit when the environment or the situation was different to what we had expected. Fitting the narration to the video and perfecting the audio track took many working hours. The more various elements are combined in video during the editing phase, the more time should be reserved for editing and finalising. After the premiere we felt quite pleased about the outcome, as it neatly summarised our understanding of context awareness. A video such as DrWhatsOn I does not need to be of perfect technical quality. We used the video to market our ideas of context awareness and to find new ideas for our future research. Even though we had tried to make the video self-explanatory, it turned out that after a few presentations and discussions we had to create a supportive PowerPoint presentation about the video and context awareness.

Turning technical and abstract visions into concrete examples is always challenging. When it succeeds, the video may have a tremendous effect on the future development and adoption of new technologies – in this project, this is something that became very clear afterwards. ∎

He doesn't find it.

5

He arrives to a grillroom in a place called Vuotso.

Provoking change

ALAN ROSENTHAL

"For me,
working in documentary implies a commitment
that one wants to change the world for the better.
That says it all."

5

Provoking change

Where as this book has so far presented ways in which the *process* of shooting and editing video can support design, this last chapter is concerned with video *as a presentation tool*. Highlight tapes and vision movies are produced with the ambition to change how people think. The first is directed inwards, towards the design team and managers inside the company. The second addresses both people inside the company and people outside: the customers.

Design is fundamentally about facilitating change: the design team may want to change products, systems and services and through this they will inevitably change the practice of the people using them. However, at the same time, an innovative design process challenges people inside the team and inside the company to reconsider their understandings and how they operate. Such changes seldom come about through rational deliberation; they require provocation and openness to discussion and reaction. Well-crafted videos have the power to provoke change.

The psychology of change

Most humans are hesitant to change their ways of doing or thinking, as change inevitably implies uncertainty about the unknown. This is as true of individuals as of groups of people and organisations. When employing video to provoke change, it is useful to understand some of the psychology behind how people react. The immediate reaction that videos trigger takes place in people's minds. However, a much greater influence can result from the social impact of a debate provoked by the video. Hence this chapter begins with a look at a theory of how change occurs within the mind and then continues by highlighting the importance of the setting around the video presentations for provoking the desired effect.

Cognitive Dissonance

The social psychologist Leon Festinger introduced the concept of *cognitive dissonance* to explain the discomfort of changing attitude (Festinger, 1957). A cognitive dissonance appears if there is such a strong contradiction between what people perceive in a situation and what they believe that they must either reject what they see or reconsider what they believe. This is a powerful trigger for change.

In a usability test at Danfoss, for instance, a mechanical engineer observed that several electricians did not mount the product under testing with enough space above to ensure proper ventilation, though he had included a drawing and the text "min. 100 mm" in the installation guide. Upon scrutinising the video recordings, it became clear that most of the electricians had in fact seen this drawing, and one of them had even used a tape measure to check that he had left enough space: 3 cm. To the usability professionals, this seemed an easy problem to solve: simply change the drawing text to "min. 10 cm". To the engineer, educated to measure everything in millimetres, it was almost painful to accept that there might be electricians who do not know precisely what a millimetre amounts to. Before he agreed to modify his drawing (*i.e.* change his belief about users), he launched into a lengthy debate about whether the invited users were in fact "real" users.

Cognitive dissonance is the discomfort of holding two conflicting thoughts in the mind at the same time. As many design problems arise out of misconceptions about how users think and act, real-world video can constitute one side of a disjunctive pair of cognitions, strong enough to challenge the viewers to reconsider their beliefs, and possibly to change their attitudes.

Unfreeze, move, and freeze

A useful concept for understanding change in organisations stems from organisational psychology: the sequence of *unfreeze, move, freeze* suggested by

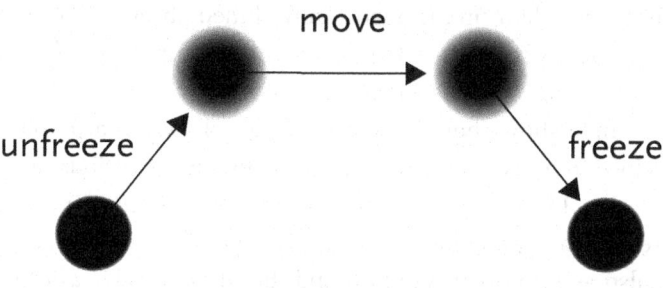

Kurt Lewin (Lewin, 1947). He claimed that groups of people exist in a comfortable state of equilibrium, in which the members have relative safety and feel a sense of control. To change to a new state (level of performance) is not a single step; change management is a process. Although each individual may diverge a little from the group standards, it is often not socially acceptable to diverge too much. Thus, it requires considerable force to "break the habit" or *unfreeze* the custom, before the organisation can *move* together to a different way of thinking and level of performance. The unfreeze phase is one where it becomes acceptable to even start talking about change. The move phase, then, is the actual transition from one level to the next. Once a change has occurred, this may not automatically lead to a new, permanent state. Group life may soon fall back into old habits, unless care is taken to *freeze* the organisation at the new level of performance.

Figure 5.1. Lewin's concept of organisational change

Yrjö Engeström (2001) presents a case in which a change process was conducted in a children's hospital. The management held the belief that the working concept of "critical paths", which they had taken into use a while ago, was working well. The idea was to describe the patient treatment as a path, where the patient is brought through different units and phases. However, on the workers' side the idea was conceived to be insensitive to patients, who had multiple problems at the same time.

A set of meetings was held to develop the practices at the hospital. In these meetings videotapes from interviews with the hospital workers were shown. At the beginning of a series of workshops the management heavily objected that the cause of the problems was the work concept of "critical paths" that they had earlier adopted. However, the combined, extensive, real-life evidence on videotapes and the presence of all the parties involved made the conflict apparent, and enabled the management to face the unhappy reality that things were not proceeding as planned. It is noteworthy that all the parties were present at the events, as this did not allow anyone to be blamed for the problems in the opportunity of their absence. The provoked debate helped to unfreeze the situation and opened up the organisation to new ideas and moves towards a better practice.

This example shows how video recordings, shown in a proper social setting, helped to fuel collaboration towards building consciousness of the unhappy reality. However, social pressure *to change* one's opinions is not the only reason to pay attention to the social setting. Social pressure *not to change* is also as important. Lewin found that it is usually easier to instigate change in individuals when they are part of a group than to attempt to change people individually. For fear of deviating from group standards, people will put up strong resistance to change when on their own (Lewin, 1947). It is thus very important to see video presentations within a broader frame of social interaction.

What opportunity does this leave for video as a change agent in organisations? Vision movies may play a role in both the "unfreeze" and "move" phases. To unfreeze, videos that pose a "what if?" question can trigger discussion in a group or organisation about the possibility of, and benefits of, changing thinking and practice. To do this, the videos need not be realistic; rather they should be bold and radical enough to instigate a reaction from the audience. Provocations to see alternative futures are helpful at this stage.

In the transition phase, when the group attempts to change its attitudes, customs, and performance, a video that shows an image of the future may serve as a vision to wish for and strive towards. Organisational change literature may call this the "management pull". Here, the video needs to depict a realistic future of use practice or company operation and illustrate a meaningful vision for the organisation's perspective.

In summary, we can think of at least three principles of how video may provoke change in an organisation: (1) as evidence of "real life" to counter prejudices against users, (2) as radical scenarios to trigger a discussion on change, and (3) as a vision towards which the organisation can strive.

● Method: Usability Highlights

*"See, how
they can't
use it!"*

A *highlight tape* is a well-known format for summarising test results in companies that employ usability testing. A highlight tape communicates findings from a series of usability test sessions with, say, four to ten users by compiling "highlights" of the most significant episodes of user behaviour. It is problem focussed, *i.e.* it shows examples of how users encounter difficulties with a user interface, or how users go about solving tasks in unexpected ways. The highlight tape is typically an appendix to a text report, and it is produced not only to convey findings, but in particular to persuade the design team or management that the test has uncovered usability issues critical enough to be dealt with before the product goes to market.

In industry, usability testing is typically subcontracted to a group of specialists external to the design team, either to a separate unit in the organisation or to a consultant contractor. The usability professionals run one or more usability labs and are skilled in organising test sessions, documenting and analysing human behaviour, recording and editing video – since video has become the dominant media employed. They are not part of the design team, because there is an expectation that a "neutral" body will produce better, "uncoloured" test results, and because the usability professionals need to maintain their specialist skills. This means, however, that there will inevitably be a hand-over process: a point where the usability professionals try to transfer "their" results to the design team. This is no easy undertaking, and this is where the highlight tape plays an important role.

A usability test almost by definition finds usability problems. To uncover problems, however, infringes on the professional pride of the designers who built the prototype being tested, and to solve such problems becomes a matter of cost and time for the team, priorities to which the usability professionals are not privy. All they can do is to present their argument as convincingly as possible and hope that the usability issues survive the priority discussions in the subsequent process.

Usability testing as a method has in particular been subject to heavy critique from the participatory design community for building on de-contextualised user tasks in unnatural environments, and for pointing out problems too late in the process and without suggesting solutions. In a sense the video card game presented in Chapter 3 was a reaction against usability testing and a suggestion to constructively develop the hand-over process into a collaborative design dialogue. However, this discussion is not the focus here; we shall concentrate on the role of user video in changing opinions and influencing design.

Persuasive videos from fixed cameras

We have seen previously that a video that triggers a cognitive dissonance may have a chance of persuading people to change their beliefs. How does one edit a persuasive highlight tape, then?

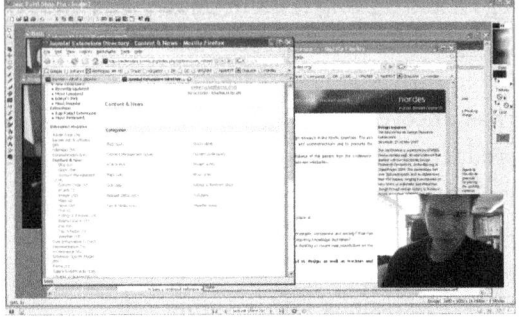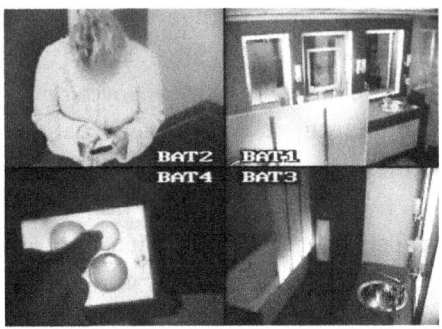

Figure 5.2
Possible split
screens at
a usability
studio

Usability labs are typically equipped with stationary cameras. This influences the possibilities to author an appealing video presentation with the material. Although the cameras are remote controlled, they are mostly set at fixed angles for each session to cover, for instance, the face and hands of the user plus the screen of the prototype equipment under test. The camera signals are recorded either as a main picture with small insert image, as a split-screen (two or four images on the same screen), or on separate video tracks to provide a choice of angles later on. Editing between fixed camera angles or split-screens without producing a monotonous, boring video is a challenge!

The camera shots in the lab are typically organised to facilitate detailed observation of human behaviour rather than narrative movie editing. Often the camera positions do not respect the "180-degree rule" of moviemaking: that all camera shots ought to stay on the same side of an imaginary axis created by two persons in dialogue or a person interacting with an artefact. Violating this rule – "crossing the line" – results in "jump cuts" where people appear to change position at edit points. Imagine the confusion if a TV transmission from a tennis match or football game suddenly mixed cameras from both sides of the field! Editing a coherent story is evidently another challenge with usability video material.

In spite of the difficult conditions in editing usability material, highlight tapes that show users struggling with company products have been reported to have a profound influence on management attitudes towards usability (Dumas and Redish, 1993). In our experience, some of the tricks successful usability professionals employ to make highlight tapes persuasive are the following:

Start on a positive note: Show things that came out successfully with the
design. Choose a few total shots to introduce the setup and test proce-

Figure 5.3
The 180-de-
gree rule
for placing
the usability
cameras

dure. This prevents discussions on whether the users may have been "misguided" in how to operate the product.

Show the face of the people in the video to enable the audience to identify with them. This makes it much more difficult to discard the people as "stray cases" who do not represent the "real users".

Pick no more than four to five usability problems of high priority. Show details and add explanatory text to make certain the audience will notice the problems. Repeat the action and use slow motion if things are difficult to make out.

Briefly show several users, if others encounter the same problems, to argue that this issue is not a "one-off case".

Discard poor quality scenes. If the scene does not have sufficient visual or audio quality to convey the usability problem, it is not worth displaying. Do not allow the viewers to shift attention away from the core message.

Make the video short and to the point. The practical limits are somewhere between 7 and 15 minutes. Remember that rather than regarding the highlights video as an objective scientific record, an audience will implicitly judge anything on a screen by moviemaking standards.

Highlight
tape
**Toons Toys
at HomeLab**
2'35"

More recent   †
affiliation:
The University
of Southern
Denmark

▶ Case story: Toons toys

Marcelle Stienstra, Philips Research[†]

The room is filled with family, friends and colleagues, and the atmosphere
is anxious. In 15 minutes my PhD committee consisting of professors and
researchers will enter the room for the official PhD defence ceremony. In this
time left before the ceremony starts I will explain to the audience what has
kept me so busy the last four years. I have prepared a PowerPoint presentation that consists of a mixture of "easy to understand" material and more
formal research results. First I explain through words and pictures the three
interactive toys that I have designed. The moment I am about to show a
video of children playing with the toys, the audience shifts into a more active mode of attention. This video always requires an introduction to what
to expect and where to look.

The audience reacts with laughter at the comments the children on-
screen make to each other, and they share the enthusiasm demonstrated
by the children. After the video I ask the audience which toy they think the
children enjoyed the most – a nice lead into the results of my study. And
again, it worked: the audience did get a good impression of how children
in general react to the toys.

Craft a careful story through the order of sequences. Break the monotony of fixed angle cameras through cuts or texts. Already before the recording there is much to achieve by carefully choosing camera positions and ensuring best picture and sound quality (see, *e.g.* the above 180-degree rule).

A somewhat different approach, rather than attempting the persuasive style, is to select the most important clips, then allow the audience to become involved in understanding what has happened, defining the problem, and discussing what could be done to solve it – like in the video card game. Provided

The Toons Toys study was part of my PhD research where I investigated the viability of two interaction design strategies, each taking a physically-active approach towards interaction design but with different perspectives on the relation between gender and technology (more about this can be read in Stienstra, 2003). The test sessions with the children took place in spring and autumn 2002 in the children's room and the study of HomeLab. This is a research facility at the Philips Research Laboratories in Eindhoven, the Netherlands, set up to observe people trying out new technologies in a natural (*i.e.* home) setting (Aarts and Marzano, 2003).

The Toons Toys study was the first to take place in HomeLab – at a time when technical staff was still busy implementing the last bits and pieces of the technical infrastructure. They regarded my study as a good test to gain experience about what kinds of issues to expect. So the atmosphere in HomeLab was quite excited. Were the toys robust enough? Would the recording equipment work? Was the observation room up to par? And of course, what would the children think of the HomeLab, the toys, the whole experience of taking part in the study?

Using video to analyse interaction. The goal of my HomeLab study was to investigate the experience of interacting with the toys from a holistic perspective. To this end I used different analysis methods: in addition to questionnaires – more commonly used within Philips – and digital logging of actions, I also wanted to closely observe how the children actually played with the toys. This required that I record each session for it to be analysed later.

the audience can allow for the time and patience this takes, it is a very powerful way of transferring findings. However, it is hard work and might come up against the "executive summary" syndrome in corporate management.

The following cases from Philips HomeLab show how video was utilised in the study of use that is triggered by completely new kinds of products. The "Toons toys" case explains how videos with fixed cameras helped to develop an understanding of how children relate to new kinds of toys. The "Bathroom lighting" case promotes the fact that, while these usability types of videos allow designers and researchers to gain detailed data on use, the videos may be even more useful in helping designers to gain inspiration

Most of the sessions took place during late afternoons and at weekends, with no other personnel present but me. I had to start and stop the recording myself. This meant running down and up the stairs of HomeLab: both the children's room and the study were located on the second floor, whilst the observation room was located on the first floor next to HomeLab. The children were left alone during this time.

I therefore had to work with camera positions that were decided upon before the session, even though the cameras were able to be remote-controlled. I pointed the cameras at the toys and one camera to capture the whole room. Each corner of each room was equipped with a camera connection point.

To balance the best possible recording of what happened when the children were playing with the toys, I decided to use two cameras per room. Each camera had a different angle on the children and toys. The technical support staff had allowed me to store one full-screen stream and one split-screen stream: more than 60 sessions of more than an hour each on HomeLab's hard disks. This required a lot of hard disk space for the streams!

The split-screen stream consisted of the recordings of the four cameras that I was using. Later, when the hard disks were needed for new projects, CD's were made of each stream. For closer analysis of the children's collaboration and their interaction with the toys, all four streams of some 20 sessions were stored on VHS tapes. VHS tapes were utilised to enable the psychology students, who did not have other equipment in their possession, to analyse the material. Although the quality of the tapes was not quite as from the visuality of the responses and interactions. Both of the HomeLab cases underline the value of video in bringing the new *experiences* of users directly in view of designers and engineers.

The case "Let's Playnt!" takes a different stance on how the highlights video was created. The video was captured with a small handheld camera, and the test was organised in the users' (*i.e.* the little children's) native environment. It displays a highly non-scientific use of video to convey the impact of the new product concept on the users. As it was also edited into an artistic presentation, it helped to move from the analysis of problems into the persuasion of others about the new potential. ■

good as that of the digitally captured material, the images were still clear enough to show the children's actions and expressions.

Convincing material. After some time, quite a few people started to wonder what all these children were doing on the research premises and what was happening in HomeLab. I decided to prepare a short presentation on my study. The most convincing way to explain what was going on was to give a "true" insight into how the children experienced the toys. To this end I chose video material from an actual session.†

The movie clip that I still use for presentations about this study shows two Dutch boys around the age of nine or ten. They display a wide range of emotions (frustration, joy, excitement), concentration and different forms of collaboration. By looking at the clip, people get an indication of how these children perceived the toys. For example, when playing with one toy, the boys are very concentrated, and hardly talk at all. In contrast, when playing with another toy they run around, jumping, hollering with laughter, whilst being very concentrated on the game at the same time. Especially this sequence always results in laughter from the audience, since the boys are very expressive in their experiences, both physically and verbally.

The expressiveness of the boys allowed many people to engage more with the story I wanted to tell, whether it be researchers at a conference or Philips management whose interest I tried to get for research into different ways of interaction with electronics. Not being able to understand exactly what the boys were saying to each other did not seem very important. ■

† Fortunately, the parents of almost all participating children had signed an informed consent form, declaring that we were allowed to use the video material of their children for research and presentation purposes.

Highlight
tape
**Interacting
with future
lights**
1'35"

▶ Case story: Bathroom lighting

Andrés Lucero, Eindhoven University of Technology
Tatiana Lashina, Philips Research Eindhoven
Elmo Diederiks, Philips Research Eindhoven

How will people experience and interact with future lighting systems in their homes? We have this question in mind as we study future home lighting with a facility of 50 or more separate light sources producing light of variable distribution, intensity and colour, which can all be controlled to create a variety of atmospheres.

Background. The project began in November 2003 and was a joint effort between Philips Research, Philips Lighting and the Eindhoven University of Technology. It was carried out in HomeLab, Philips' research lab in Eindhoven. Earlier studies indicated that people invest substantial effort into creating appealing ambiences in their homes especially with light, and suggested that the bathroom is one of the appropriate locations to do so. In this project we wanted to explore what kind of ambience would enrich the daily rituals people have in their bathrooms and what easy-to-use solutions we could offer them to compose and control ambience.

User study and design. Our first task was to conduct a contextual study on how people use their bathrooms and identify any needs that people may have concerning lighting in this context. We designed Cultural Probes which consisted of a diary and a disposable camera that allowed the research team to enter and study a very private place in the home such as the bathroom. Through the information and the pictures that our participants shared with us we discovered a diverse, and in some aspects unexpected, view of how people use their bathrooms.

Based on what we learnt from the Cultural Probes we designed several interaction concepts for lighting in the bathroom that aimed at reducing the complexity of interaction with such a rich lighting system. A final interaction concept was defined through a number of iterative cycles. It comprised a user interface with an abstract representation of the bathroom and ambience elements (represented by natural phenomena metaphors such as a sunset, a cloudy sky, a lavender field, an ocean, and so on) that could be combined to create different ambiences.

Usability test. We invited people to evaluate our design in HomeLab. The lab is built to resemble a real home and is equipped with discreet cameras and microphones. This has the added value of allowing us to observe how our participants experience and interact with the prototype while participants feel they are in a natural home environment.

We asked participants to complete two sets of tasks in order to test the usability of our design. The first set of tasks consisted of short everyday activities that people usually perform in their bathrooms (*i.e.* switching the lights on and off), while the second was connected to a new feature of the system that users would only perform sporadically (*i.e.* creating atmosphere for relaxation). At the end of each set of tasks, we asked our participants to evaluate different aspects of interacting with the system by using the attitude scales of the Technology Acceptance Model. To evaluate their understanding of how the system worked and the metaphors used we also asked the participants to describe the interaction as they understood it and the meaning of the different user interface elements.

From the control room, we could direct the four remote-controlled cameras available in the bathroom of the test lab. We wanted to keep the cameras in locked position, angle and zoom level to ensure capturing similar data across all participants. However, in order to properly capture the spontaneous facial expressions of our participants we had to make some small adjustments with one camera while the test was in progress.

Video to compare the usability test results. Initially we thought the videos would allow us to go back and have a detailed look at the physical actions (*i.e.* number of times the switch was pressed) and the time it took participants to complete a given task. Additionally, we wanted to have lasting evidence of some of the comments and reactions from participants while interacting with the system. The video would allow us to look back and find the reasons behind some of the difficulties participants encountered with a given task. Having the video records we could compare them with the mean ratings from the TAM questionnaire and confirm or reject some of the results of the usability test. In this way video would help us overcome a possible "experimenter bias" that might occur when participants try to be polite towards the evaluator and creator of the system by giving higher ratings. We could analyse the video data and see whether the system was as easy as the users reported in the questionnaires.

Video to capture and communicate experiences. Very quickly after we started the first couple of evaluations we discovered new aspects to using video that we had not previously taken into consideration. For example, with video we were able to see how people experience the system by paying attention to their non-verbal communication (*i.e.* facial expressions and body language), and how they reacted to the lighting changes triggered by their interaction with the system. Above all, by showing some of these reactions we had a very convincing and clear way to communicate our test results. Results of a usability test are usually presented through statistical data and graphs. It can sometimes be difficult for some viewers to see clearly through the data and read the true meaning "between the lines". However, a one-minute video with real experiences of users interacting with these systems appears to be very powerful to illustrate the findings in addition to the data obtained from the tests.

Video probes. Looking back at our process, we realised that Cultural Probes can also benefit from the use of video. Video is a richer medium than a still image and may thus become a better alternative to the photo capturing commonly used in Cultural Probes, especially since it has become much more affordable. A video probe or diary consisting of a digital camera may allow participants to

Moving organisations

In interaction design, vision movies are videos that show what interacting with computers can be like in the future in movie action style. Apple Computer's "Knowledge Navigator" (Dubberly and Mitch, 1987), Hewlett-Packard's "Imagine" (1992), and Sun Microsystems' "Starfire" (1994) are examples of large corporations producing vision movies on a professional moviemaking budget. These movies have a duration of around ten minutes, and they show how computer systems and services may be integrated in a future five to ten years ahead.

"Knowledge Navigator" shows a university professor using a portable computer (with foldable screen and built-in video camera) to organise a presentation on deforestation in the Amazon rainforest. He has a videophone conversation with a colleague to effortlessly compare and exchange visual data. The professor interacts with the computer in natural language through a software agent, depicted as a helpful butler in the corner of the screen.

In "Starfire", a product manager of an automobile manufacturer suddenly finds herself challenged to defend the market introduction of her new sports car model against the plans of another department. Within hours she needs to compile a convincing board presentation, using a large curved desktop display to compile a variety of data from around the world. However, instead of ending here, "things must go terribly, terribly wrong" in the movie, according to director Bruce Tognazzini (1994): at the (tele) board

make several short videos capturing some of the experiences and stories they want to share with us while working with the probes. This would result in richer data, contributing to a better understanding of people's needs and motivations.

It goes without saying that video is very useful in usability evaluations; it is a valuable medium for capturing user experiences with a system. In our experience we learned that video is a valuable tool that would also be beneficial in other stages of user-centred design, particularly in providing more contextual information. It has the potential to provide a deeper view on people's attitudes, needs and motivations by capturing, for instance, non-verbal forms of communication. ∎

meeting she is attacked on facts by her opponent and needs to counter the arguments by retrieving new data in a split second – which the computer system luckily helps her accomplish.

In drama theory, the key ingredient to gripping drama is conflict. In "Starfire" the conflict is played out between the two division managers with computer weapons. Building suspense in movie production is highly challenging: both the actor's performance and the credibility of the causes of the conflict influence how well the excitement builds up. Moreover, the audience needs to accept the way the context, the characters and their roles are presented.

"Starfire" opens with an airport scene with a landing airplane, passengers stepping off, and then the startling message that "your personal office is just around the corner" triggers the curiosity of the protagonist. The beginning of a story sets the stage for understanding the rest, and thus sets the audience's expectations. This is an important element of a movie, because, as Brenda Laurel (1993) says: "When we have no particular expectations, discovering new information is a simple and relatively unremarkable experience." In "Starfire" it is rather the climax that has the strongest impact on the experience, and it is when the understanding of the meaning of the story develops most.

Drama directors and scriptwriters are trained to create stories that capture people's attention. They are highly skilled in developing lively detail and structural coherence into a play. The German critic and playwright Gustav Freytag introduced his famous triangle in 1863 to explain the development of dramatic tension, or suspense, in a play (in Laurel, 1993). The model is based on Aristotle's concepts of "complication" and "dénouement": it shows *rising* and *falling* action. Rising action leads to a climax or turning point; the falling action is everything that happens after the climax (Figure 5.4).

Figure 5.4
Freytag's triangle of dramatic tension

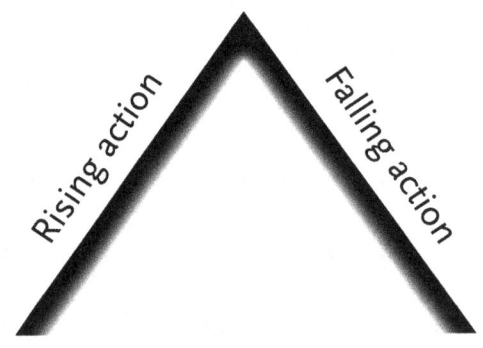

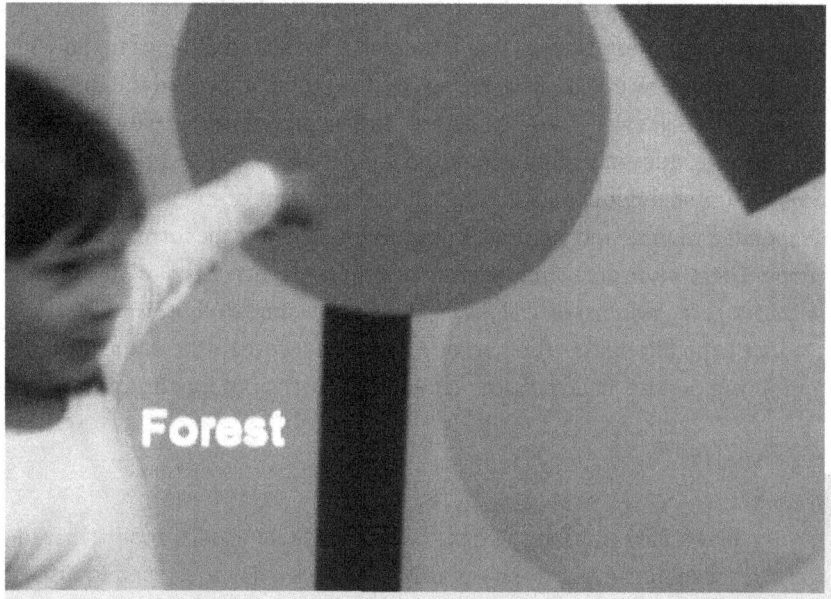

Forest

▶ Case story: Let's Playnt!

Pia Salmi, University of Art and Design Helsinki

Highlight
tape
Let's Playnt!
1'49"

We show our 1920's style video in the final presentation of the User Inspired Design project. Everybody is laughing, including the teachers who are supposed to criticise our ideas. The video works so well that both the teachers as well as the fellow students are "buying" our idea in the closing evaluation.

The user inspired design course (introduced in Chapter 3 in the case story "Conceptual door") aimed to educate us about user-centred concept design. We explored children's (three to six years old) communication to gain new insights into a conception of "door". We had reframed the idea of door, and understood it as the dialogue between children – as a door to another's mind. It made us look at situations where children were negotiating with each other as well as situations where they did not.

We had two main observations: first, we noticed that small toys made the children focus on their individual rather than collaborative play, and second, we observed that when the children played with abstract shapes, such as pillows, they discussed a lot about what the pillow is in play. We took these ideas into our Playnt concept.

The dimensions of the triangle are *time* on the horizontal axis and *complication* on the vertical axis. Fundamentally, the triangle shows that during rising action new questions arise, *e.g.* "Why did the man steal the car?" The falling action answers these questions. In real plays, the dramatic incidents usually introduce many questions and answer some at the same time. However, the model delineates the overall development of the action, which is elaborated in more recent models (Figure 5.5) that take the form of a curve or slope. These more detailed models include phases such as *exposition, inciting incident, rising action, crisis, climax, falling action, and dénouement.*

The exposition (a) is the part of the play that reveals the context for the unfolding action. It introduces characters, environments and situations.

The Playnt case was based on the ideas of abstract and big shapes that trigger the collaborative negotiation of play. When the children needed to discuss how the pieces were related and interpreted, they were expected to engage in discussions with each other. Moreover, we did not want to set any specific rules for playing, which – we thought – would also enable the children to explore more easily new possibilities with the things they were provided with.

The name of the concept comes from the words "paint" and "play". Playnts are large, colourful, two-dimensional shapes – pieces of paint, which came in "Playnt buckets". They are self-adhesive; they can be attached to walls, floors, furniture, *etc.* They can be used for creating art works, constructing landscapes and making backgrounds for playing.

Playnting with a bucket full of Playnt. We tested the concept with Maija (5) and Roosa (4), and we were quite surprised at how well the concept worked. The girls had never seen these kinds of self-adhesive pieces of paper before. After initial hesitation nothing could keep the children from trying out new ways of using the pieces of paint. They spoke aloud and explained what they were doing. The older one guided and the younger followed.

> – *Can we really attach these on the wall? Great!*
> – *This is a lamp*
> – *Here is a table under the lamp*
> – *Here is a tree, and here and here.*

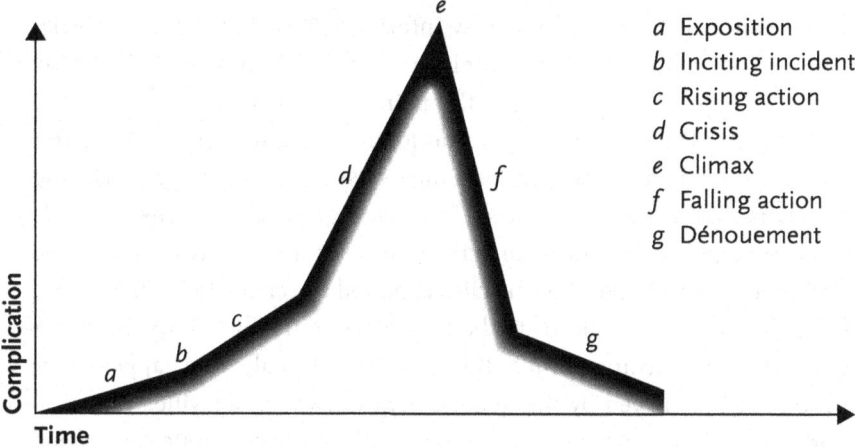

a Exposition
b Inciting incident
c Rising action
d Crisis
e Climax
f Falling action
g Dénouement

Figure 5.5
The modern shape of dramatic action (adapted from Laurel, 1993)

This went on until the whole wall was illustrated with homey objects. The next task was to make something on the floor. "Yee, let's make a track where we can play." The girls ran through the track, jumping on yellow, walking on red and jumping over brown pieces. One child hid small pieces under big ones and the other one had to find the hidden pieces after the first girl's hints. They also had a running contest on the track.

We wanted to show how well our concept functioned, and we thought video would be the best means for this, as we had the material on tape. We had about 30 minutes of video from our test. We watched the material quickly and showed it to Maija's mother. We used the tiny camera screen with fast forward to observe it quickly. It looked quite amusing in high speed, and this might be the reason why we decided to use it also in our final presentation.

Deciding to use the double speed made it difficult to use the original sound. We did not find it appropriate to set the original soundtrack to the double speed movements; instead, we took Irving Berlin's song "Alexander's Ragtime Band" and put it on the background. The feeling was perfect!

We then picked key lines of the girls' speech and put them on the final clip as texts. The editing took a couple of hours and resulted in a piece that lasts approximately a minute and a half. According to the feedback by the tutors of the course the video was conceived as the most memorable of all materials presented during the course. We still think that there cannot be a better way to convey the value of an idea! ■

Exposition as the revelation of new information, or perceptible material, continues throughout the play, but its importance diminishes towards the end. The inciting incident (b) is the point in the play where the central action starts. During the rising action (c) the characters pursue their central goals, formulate new plans and meet obstacles along the way. During the crisis (d) the activities proceed faster and grow more intense towards the climax (e), which turns all activity towards a single necessary direction. At the turning point in the climax period the characters either attain their goals or fail. This leads to the falling action (f) that shows the consequences of the turning point to the characters. Usually this happens very quickly. It is followed by the dénouement (g), where activities return to normalcy.† This structure is also recognisable in the corporate vision movies described above.

The French word dénouement means "untying" or "unravelling". †

Where both the Apple and Sun vision videos were rather stationary acts of one or two actors in office-like sets, Hewlett-Packard's "Imagine... A Vision of Health Care in 1997" was a much more complex production with suspenseful cross editing between three parallel storylines set in a large hospital. The hospitalisation of an acute heart patient, the medical diagnosis of a little girl with mushroom poisoning, and a hospital management's struggle to maintain funding for the hospital. Also here, the accessibility and integration of the right information at the right moment helps the protagonists to a successful resolution of their conflicts.

These vision movies are prototypes in the sense that they demonstrate ideas about technology in a rich social context, without the devices and computing having been developed to a stage where they actually work. They communicate through a compelling story that allows the viewers to themselves imagine what it would feel like to own and interact with such technology. To create this ultimate illusion vision movies employ special effects as used in science fiction movies: blue-screen recording, superimposed computer animation, *etc.* The result can be very convincing indeed.

Did these vision movies have an impact, then? Apple's "Knowledge Navigator" received widespread attention as a statement about the future of computing and it helped pave the way for the Apple Newton – the first handheld computer of its kind.

These videos can do wonders for the company's image. Apple's Knowledge Navigator positioned Apple as the "futures company" for many, many years. (Tognazzini in Bergman et al., 2004).

In human computer interaction circles the video even provoked tense discussions about just how far in the direction of human-like agents computers will be able to develop – and whether this is actually a desirable and ethical track to follow. "Starfire" was explicitly aimed at a strategic change in how people in the Sun organisation saw the future, and HP went to great care to ensure that all medical procedures in "Imagine" were correct to make the vision appealing to the medical community.

● Method: Vision Movies

Vision movies are video presentations that communicate design visions; they convey a message as efficiently and compellingly as possible – the message being the design concept and its value in use. They may be developed in the manner of full-featured movie productions, or be crafted with a small but skilful team of enthusiasts. The aims dictate how the stage needs to be set, and how high a professional quality is needed for the purpose.

"This is our future"

The first condition in producing a convincing vision movie is that the team has a powerful design concept. In this way working with vision movies is quite different from the scenario design techniques described in the previous chapter. In those examples, the point in scenario design about products was to develop the designs *while* improvising their future use. Developing the designs along with the video production in vision movies can, however, be downright dangerous. According to Bruce Tognazzini, the producer of "Starfire", the ease with which moviemaking builds illusions will tempt designers away from the possible towards fantasy, including technology that may not be available for another one hundred years, rather than within the ten-year target (Tognazzini, 1994).

Though Tognazzini argues strongly for the vision movie to present a real proposed system, one that can arguably be realised within a limited span of years, his is not the only opinion. At a CHI conference panel Eric Bergman, also from Sun Microsystems, claimed that:

> ... *vision videos are important, but not because they suggest a true vision of the future. They are important because they inspire us to think about what might be. If we can value them for that inspiration, and not for the specifics of the vision, then we have tapped into their true value.*
> (Bergman et. al., 2004)

Vision movie
It's UI Love
Video
8'27"

▶ Case story: It's UI Love

Case author: Urpo Tuomela, Nokia Corporation
In collaboration with the University of Lapland

The movie is planned as the 76th episode of an imaginary and famous soap opera "Ubiquitous Women" – a drama taking place in Paris aka New Hong Kong in the year 2011. In this episode John, a corporate employee, has problems with his boss Louise, who simply falls for him. At the same time John's new girlfriend Katya eagerly wants spend more time with him. The video shows how they go about solving their personal liaisons with the help of new and perhaps "not so new" technology. Some privacy and security problems are also addressed.

The It's UI Love project at the University of Lapland was related to our mission in Nokia to anticipate the next big wave, *e.g.* to find and demonstrate

Bergman also criticised most vision movies seen so far, in that they are not actually user-centred: they present technology enwrapped in some imaginary use rather than realistic work practices enhanced with technology ("It is easier to turn on the light using the wall switch instead of my mobile phone"). In short, a credible design concept that is based on credible accounts of future trends as well as relevant discoveries into user practices is a pre-condition for producing a grounded and convincing vision movie. Moreover, the vision needs to be open enough to allow for the imagination of the audience to take off. Lindholm and Keinonen (2003, p. 142) outlined the role and proper character of visions:

> *In introducing a vision, it is not important to paint the complete picture.*
> *A vision has to leave ample room for imagination. If there is nothing to*

technologies that might have promising product and business cases. In 1997 this quest was also set forth for technologies of ubiquitous and wearable computing, and the vision video "It's UI Love" was an attempt to visualise the technical and non-technical aspects of future communication: What does ubiquitous computing actually look like? How can we make a set of networked devices that work in the background and form a seamlessly operating intelligence? What are personal communication devices, and what is their role? Finally, if all the computing and intelligence disappears in the background, how can we command something that is invisible?

Our research team at Nokia had created technologies and demonstrators for wearable computers, home appliances and context aware services, which could enable new means for communicating, producing and presenting information. However, we needed to take these ideas further and visualise them. The social impacts of these technologies with privacy and security issues were also untouched and needed to be researched.

We saw video animation as a possible medium for modelling and presenting our ideas. With video we would be able to raise questions about technological development and possible directions, and to discuss the user experience with new kinds of user interfaces and product concepts. To do this we felt we would need to look ten years into the future, when we could assume that network capacity is infinite, the intelligence in devices practically costless, and all devices are able to talk to each other.

spark the imagination, how can one get the audience excited? The vision has to provide the spine, the goal or beacon to guide the design team throughout the development process.

The second condition is a compelling storyline. Movies live by their story. A rosy red story showing people creating perfect things with their perfect devices makes a rather boring movie. In real life nothing runs smoothly. Things go wrong, people misunderstand each other, dilemmas arise, and people need to react on the spur of the moment.

> *Film is a powerful medium, capable of either showing perfection, thereby stifling discussion, or showing imperfection, thereby promoting debate.*

We established collaboration with the Faculty of Art and Design at the University of Lapland to work on such a vision video as a start to the project Ubilink in January 1998.

Planning. To minimise risks the project was divided into planning and production phases. The required amount of work would raise the costs very high, so a solid plan of the activities and costs was necessary.

The planning started with writing the initial script and modelling wearable computers, future clothing, a smart home interior and various home appliances. Already in this phase the challenge showed its character: it was very difficult to visualise a convincing future lifestyle with credible persons living in a completely new kind of environment. The new interaction methods and "invisible technology" also caused extra contemplation.

The plans, including design posters, design mock-ups, script, schedule and resource needs, and the plans of video recording locations, used tools and cost estimates, were reviewed in Nokia in May 1998. The decision was not an easy one at Nokia. From the company side we were pleased with the design work, and the plans gave rather realistic estimates for the required work. However, the estimated costs were very high, and there were risks related to schedule, 3D model-, animation- and composition technologies. It was clear to everybody that we were now planning to produce a video on a scale that had not yet been done at the University of Lapland.

In building a video prototype, we felt an ethical imperative to show the limitations of our designs. (Tognazzini, 1994)

To write a story like this is much more demanding, not only in terms of scriptwriting competence, but also for the design concept. This will set severe challenges to what devices can do and how people are able to interact. In short, scriptwrite for conflict and imperfection!

The third condition is high-quality video production: convincing acting, directing, shooting, and editing. Vision movies require a sufficient standard of acting and movie production to allow viewers to concentrate on the message. To be regarded as a serious argument, the vision movie needs to respect the same expectations we all have when turning on the television. In the case of

Video production. The 3D models of home interiors and furniture had to be designed starting from vague ideas and visions. The work required that an entire apartment be invented and modelled. At the same time, a manuscript with a compelling story had to be written. This parallelism soon led to a traditional chicken–egg problem: how can we create a story if we do not know the devices or the environment where everything happens, and how can we design products and environments if we do not know the people using them or how they relate to the story?

Finally we had everything ready on time. The video shooting took place in the vocational college Länsi-Lapin ammatti-instituutti in Tornio, which featured a suitable studio with skilled assisting personnel. The editing lasted longer than was originally planned due to the time needed for tuning the 3D models, for example, to adjust lighting and textures in interiors and devices. In spite of the slight delays, the video was ready at the end of October.

"It's UI Love" has had an impact on numerous projects. It has been presented in various situations at Nokia and also at numerous seminars and conferences, and the two other case stories in this book about the context aware phone were initiated by the making of this vision movie. With the cost required for realising virtual 3D-models and animations, it is clear that this is only appropriate for long-term visions at corporate level. For visualising short-term research ideas, *e.g.* of less than five years, one would need cheaper and faster production and to focus on the business benefits.

vision movies, poor shooting, acting, and storyline cannot be hidden behind the real-world credibility of field recordings or the happy sketching quality of improvised design scenarios.

Novices to video often lack the understanding of the basic concepts in story telling, of the visual language of video, of the rhythm of editing *etc*. Even experienced video producers may create results that are too long and too boring to watch. To produce a convincing vision movie some measure of professional assistance is recommendable – unless one chooses to experiment with a simplified, sketchy style, as shown in the last case story in this chapter. Yet, a good design concept and a compelling storyline are still preconditions to making it work.

We also found that this project was a little too big and too long for university students. Even if this kind of co-production with students is interesting, it should last only one semester, and it should have a clearer focus. The students in the core team worked as full-time resources all during the summer months, but during spring and autumn periods they had other studies in parallel. They also came from different departments with their own interests and working in this kind of large and multidisciplinary project was new and challenging. This also explains in part why the project lasted so long.

There are few techniques that can be used for small-scale modelling of designs targeted to the not-so-distant future. One can dramatise experiences with existing or imaginary prototypes and user interfaces and capture these with video. However, video is then used mainly as a design tool and not for marketing or promoting corporate level visions. I think that any artistic work with multi-disciplinary teams and open specifications is an interesting process and has the potential to create something new. Thus, this kind of futuristic video should be produced every three to five years to see how people see future technology and life changing and developing.

Some people at Nokia would have expected a sharper focus on technical issues over the social aspects. However, when thinking about how technology shapes our life and social relationships, the dilemma is still there: is there any better way to visualise and present future lifestyles than a video animation with an interesting story?

– I'm still hoping to see the next episode of this famous soap opera! ■

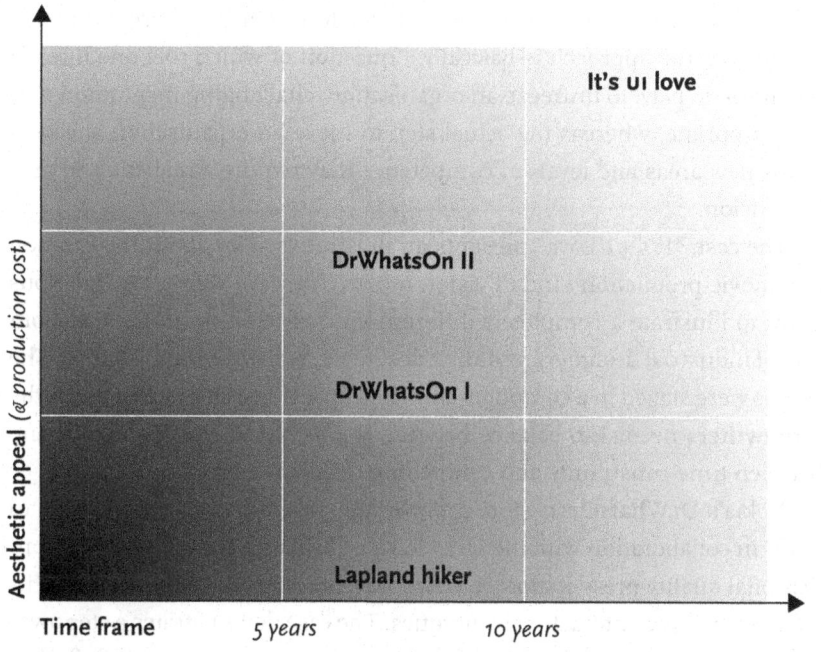

Figure 5.6
The presented case stories in relation to the effort on visual appearance and relative time frame

The fourth condition is appropriate balance. In vision movies the production team needs to find a balanced focus between movie aesthetics, storyline, and the concept. The longer the time range, the easier it is to focus on making a compelling story. The closer to the actual development process and, hence, the more specific knowledge the developers need about the technical details of the product, the more interested the audience becomes in those details. In such a situation the crisis of the hero who saves his girl from a fire, might lead the focus outside the interest of the relevant audience. The case stories' relation to time range and aesthetic appeal is presented in Figure 5.6.

To create the ultimate illusion of future prototypes vision videos such as "Starfire" and "It's UI Love" employ special effects as used in science fiction movies, such as blue-screen recording and computer animation. However, the production of these effects requires heavy investment. The "Context aware mobile" case (Chapter 4) describes a contrasting way to create a vision movie. The video focussed on exploring the opportunities for a context sensitive phone and it was visually very rough. However, it developed a nice story conveying the impact of the design on the life of the protagonist.

To return to the unfreeze, move, freeze terminology of organisational psychology, the approach is basically a question of which role one intends the movie to play: to unfreeze an organisation, challenging inspiration may be appropriate, whereas the actual step to move an organisation ahead towards new areas and levels of competence may require a realistic and gripping vision.

The case "It's UI Love" presents an illustrative story about the creation of a movie-production kind of vision movie. The case shows an ambitious effort to illustrate a completely different kind of reality with the collaboration of industrial designers, costume designers, real actors, and studios. The scenes were staged in a computer-animated space. The story reveals how the scriptwriters needed to balance between the technical and social foci, and between time constraints and compelling results.

Nokia's DrWhatsOn II story grew on the experience of making "It's UI Love" in collaboration with the University of Lapland. It displays a high professional quality presentation of a complete product concept illustrating its features and new interaction techniques. The case demonstrates a step away from storytelling towards the use of acting as a backdrop for the info flashes conveying the details of the product concept. Hence it is a wonderful example of how challenging and time-consuming it is to ensemble a complete vision movie also with a good story.

Where the two cases mentioned above provide examples of high-end vision movies that were crafted with many professionals, the "Lapland hiker" case displays a contrast. It shows that effective vision movies do not necessarily require a movie-production budget. Moreover, it accommodates a great variety of the methods and theories that have been introduced in this book. For example, it shows how Boal's theory of Image Theatre (create real image, ideal image and image of the possible transition) and Freytag's theory on movie structure (Freytag's curve) are employed in a practical case that helped to develop new concepts for knowledge management. Semi-improvised acting by the real users supplied details about a user-practice while also displaying the co-designed opportunity for development. The workers represented themselves, which makes the video highly credible and loyal to the real interactions that the workers will encounter in their work in the future. ■

▶ Case story: DrWhatsOn II

Case author: Urpo Tuomela, Nokia Corporation

Vision movie
DrWhatson II
6'18"

– Info-flashes?

   – Yes, we will use various information flashes with some animations to describe how the user interface concept works. Not a single word will be spoken!

Background. We started the DrWhatsOn project in January 2000, and the focus of the project was to create a concept for a context aware mobile device for people in office environments. We concentrated on user interface related issues, for example, in what context awareness provides to end users in terms of applications and user interaction. In the beginning we worked to study users' preferences concerning a few concept alternatives and explored the contexts and tasks of potential users. After the validation work we started developing selected applications and their user interface designs.

   Soon we understood how the product concept was a special case. The context awareness technology had not been used in mobile devices' user interfaces before. Hence the concept might turn into a trendsetter for a whole style of communication. This posed an extra challenge to the project team: Could we really shape the direction of future personal and context aware communication by creating the UI concept and a related video showing its unique and remarkable user experience?

In May 2000 the project had proceeded to a phase where the user interface concept was nearly ready. Once the user interaction team had designed the logic for the user interactions, we had enough material to start planning the concept video. We also had a few design mock-ups of the DrWhatsOn product concept, created by an industrial designer. The models were non-functional, so we had to imagine how to express the features in a way that would be illustrative enough.

We started planning this second video soon after the first DrWhatsOn video had had its premiere in May. The purpose of the first video was to promote the idea of context awareness in everyday situations. This second video focussed on the user interface concept in more detail. At the same time as the level of detail increased, the video-making team grew from the previous project to include a graphic designer, two interaction designers and one industrial designer.

The idea and the script. We began the work by clarifying the purpose of the video and outlining the target audience. This was not difficult, because just like the project itself, the video was to be used inside Nokia to promote new applications and user interaction techniques for context aware mobile devices in the office environment. We would illustrate both the features of the applications and the novel interaction techniques. It would also help to sell the ideas to managers, who could then take them into products. As a piece of art, we aimed for it to be remembered long afterwards and be spread widely within Nokia.

Soon after the kick-off meeting we found ourselves in a passionate hunt for a good story. Brainstorming sessions were held to gather ideas and proposals. The issues included such topics as personal security and identity, automation and non-intrusive user interfaces. The ideas varied from drama to science fiction, but the essence of the story, as we conceived it – the humour and capacity to capture all important aspects of the concept as a whole – were still missing.

The DrWhatsOn product concept was targeted to corporate employees working in office environments. This gave us the context of the story and made us seek interesting situations in our own everyday office experience. One of them seemed to suit our needs well. We decided to focus on the survival of a new

worker on her first day at the new office with the help of the DrWhatsOn device.
The idea was written into a synopsis:

"The video shows how a newcomer survives her first moments/days in the company with the help of DrWhatsOn. The newcomer gets her brand-new DrWhatsOn device from her boss with a very short initiation session. The rest of the video follows her experiences during the day with the new device in a positive, humorous way. At the same time her old high school friend is working in the company IT department as a system administrator. A couple of interesting messages are sent to her, and finally they meet each other after many years."

The whole video team was involved in composing the script. Together we planned each scene and the applications and interactions that would be shown. The situations were to illustrate the features, such as providing identification in order to gain access to the building, providing help to find a meeting room, and using hand gestures to interact with the device. The story appeared to be quite demanding in terms of acting. A great variety of situations were included with some bits of dialogue. When the script contained 20 scenes we stopped.

It was time for a serious evaluation: Are we able to achieve this? Does this fulfil the requirements that we have set for the video? The answer for the latter question was "yes". But for the first question it was "maybe not". Our approach would need professional actors with fluently spoken English, and the final video would extend to a length of over twenty minutes. It was definitely too long. So, we re-structured the script and reframed our intentions. We turned back to the purpose of the video: What are we really trying to say with this video?

Soon we discovered the simple answer: the video would be about the user interface concept. With this insight we began to explore opportunities to illustrate the concept without the need to invest extensive resources in the development of the story and in hiring professional actors. The discussions resulted in the idea of presenting the key features of the concept as information flashes on the screen. The acting would not contain any dialogue, and it would almost fade into the background. This solution helped to cut down the number of scenes that would have explained the features. Moreover, it helped to reduce the number of actors.

The final version of the re-written manuscript contained nine scenes. We threw away the original story and kept only a few allusive episodes promoting the corporate work context. The planning phase of the video took much more time than we originally anticipated. We were behind schedule, but fortunately the videotaping went without a hitch. We used the new Nokia House in Espoo as the physical setting for the scenes. Two cameras were used in five scenes and additional lights in two. We videotaped everything in one day.

Info-flashes. Once the script was ready, we started to design the info-flashes. We discussed extensively how to make them both sellable and self-explanatory. After a few iterations we decided to use three fields: the uppermost field explained the situation, the middle section displayed the device with interaction animations, and the bottom explained the UI concept. The video would stop at the info-flash at a chosen moment in each scene. A lot of work was spent on the visual aesthetics. The timing and rhythm seemed to be surprisingly important for making them work as intended.

The making of the info-flashes was helpful for crystallising the value of the product concept into these marketing-type texts. These were utilised in the info-flashes. Moreover, during the discussions we found the slogan for the concept: "DrWhatsOn, my new dimension of communication". This became the main selling idea of the whole concept.

An initial cut of the video was ready as we started adding the info-flashes. Then we realised that the music was a problem. Petri and Schbert, who were responsible for the music in the previous video, were working on other projects and did not have time to compose or finalise the music for this video. However, Schbert had given us a piece of music that he had composed a couple of years earlier. Unfortunately, it was not optimal for this video, but we had no choice. The video was completed by editing the info flashes into their places. With the music and background acting the info-flashes seemed to function rather well. We were happy, even though the premiere was held more than a month behind schedule.

Experiences. The "new dimension of communication" became the advertising slogan to promote context awareness and its possibilities for enhancing personal mobile communication at Nokia. DrWhatsOn was a first-of-its-kind concept, and the video communicated a concrete vision. It displayed illustrative guidelines for creating new context aware solutions, and it served the planning of later projects. Several subsequent projects have been launched to work on the related technical issues to facilitate the realisation of the vision.

The first DrWhatsOn video relied heavily on the story. However, in this later production the story faded into the background. A good story has value, but it is important to remember to present the technical solutions and concepts informatively when the video aims to concretise visions. When the audience consists of designers and engineers, concrete information on the technical opportunities gains value over an interesting story. A gripping story might have made the concept more credible, but this time authoring a great story was clearly too demanding within the available resources. Focusing on clarifying the key value of the product concept was definitely a rewarding activity, and it forced us to crystallise what DrWhatsOn was really about.

Video production needs a skilled team, and the production easily takes more time than creating a PowerPoint presentation. However, when it comes to concept presentations, I believe that good videos can last longer than any slide or Flash show. Vision videos can really shape the direction of our future. ■

▶ Case story: Lapland hiker

Salu Ylirisku, University of Art and Design Helsinki

Vision movie
The novice scenario
3'56"

Suddenly I realise that I have to play the role of the bartender. I am operating the camera and pointing it towards the first worker, who is acting out the phone call with another worker, in turn acting the role of a hiker who has lost his wallet somewhere in the forest in the hinterlands of Finnish Lapland. The part where the hiker asks the bartender for the location of the nearest bank branch will come soon. How should I talk in such a way that I sound like a "real" Laplander working at a local bar and grill?

The project "Luotain – design for user experience" was a four-year project at the University of Art and Design Helsinki, which aimed at developing user-centred processes and methods for product concept design. The case with the cooperative bank Osuuspankkikeskus was one of seven case studies conducted in the project, and it aimed to develop new insights into what knowledge management is, and the kinds of opportunities there would be for development in future. The case also aimed to construct high-level design drivers for creating fluent knowledge management in the banking context.

The project was conducted in autumn 2002. It started in September with a broad focus on knowledge management. In the early negotiations with the bank's representatives we decided to focus on the telephone banking service, which was considered the most challenging area from the knowledge management point of view. After the contextual study, the material was briefly interpreted to find the core challenges to knowledge management from the use situation point of view. The work in telephone banking includes significant real-time informational requirements combined with an extremely wide variety of situations that may involve knowledge management; a close understanding of the situations was therefore crucial to developing solutions that fit.

User scenario workshops. The findings were presented to the participants in the first of three workshops, which aimed at developing video scenarios. The workers were the same as those who participated in the contextual study. They included men and women, young and ageing, with focused expertise or broad experience, all the major roles of employees in the phone service, and with long and short working experience at the bank.

After initial discussions we decided to create four scenario stories about one particular challenging situation, which was focused on in detail. The first story described how the experienced worker handles the situation successfully. The second presented how the novice fails to deliver decent service due to a lack of knowledge. The third scenario explored an ideal situation, which was further

developed into a realistic future situation with appropriate knowledge management solutions in the fourth scenario.

We began to build the scenarios based on four abstract criteria for a challenging
situation in terms of knowledge management. The first criterion was simply "a difficult situation", which immediately triggered discussion about a caller who has lost
his identification and money, and is travelling away from his or her home district.
We refined the situation as I presented the other three criteria, which were time pressure, the need for specific details, and the need for newly updated information.

Based on these criteria, we began to write the script for the scenarios. I kept
asking questions like, "Who is the caller?" "Where is he?" "What happens next?"
"What then?" and "How does this happen?" The participants themselves wrote
the script. In the first workshop we wrote the scripts for the novice and the expert scenarios. We discussed who would take each of the roles, which was quite
a fun and at the same time a sensitive discussion. We did not want to disturb
the real phone service work, so we went to another room that resembled the
real phone service environment to record the scenarios.

We planned where each of us would stand, or sit, during the shooting, and
roughly went once through the plot to memorise the key points. Then we started
to capture the action in an improvised manner based on the rough plot. We did
not decide what the workers would say, but rather had the overall scheme of
the action visible in front of the workers. They were professional phone service
workers and were used to improvising on the phone in the course of their daily
work. We shot each scene twice from different angles to increase the dynamics
in the frame when editing the final scenarios.

In the first workshop we decided that we could use drawn images for the
part where the caller is presented in the bar and grill in Lapland. Then we would
edit a voiceover to explain what was happening in the scene. We began the
second workshop by recording this explanation that one of the phone service
workers read aloud.

After we had also filmed the novice scenario, we went through the plot in
detail to explore the places where knowledge is needed. Based on this, we began
to create a story about the ideal situation, where the worker knows everything
at the right time and tells it to the caller in an encouraging fashion. Despite
having already gone through the plot several times, we still found some new
information relating to the handling of the lost credit card that should be told
to the caller. We then went to capture the ideal story on video.

In the third workshop, we discussed how the situation may be handled with future solutions. After considering several options to accessing appropriate information, we went to capture the last scenario, in which the same novice played the role where he is able to serve the caller successfully. At the end of the third workshop we also discussed what the design drivers of future knowledge management should be. For ensuring that the design drivers as well as our solution would also be useful in other situations, we discussed two radically different situations encountered in the contextual study that may be difficult in knowledge management.

Then I started editing the scenarios, drawing pictures, and tuning the voice of the person on the phone with sound editing software to create a phone-like impression. Editing was slow, since I had to learn new editing software that was not known for its usability. Editing took in total 14 working days. The scenarios were shown in January in a session with the participating workers as well as several people from the bank's management.

Involving real users can be more efficient than doing things otherwise. It took only a couple of seconds for the phone service workers to identify a difficult situation when they were asked to do so in the workshop. I (as an external person) would have needed to ponder it much longer – and it is likely that I would not have discovered as intriguing and lively an example. One certain, great advantage to creating scenarios with real workers is therefore its efficiency and effectiveness. It was quite fun as well.

The method enabled us to link abstract themes and desires into concrete practice. After the scenario show, one of the managers commented that he had never before realised how much a real phone service situation requires the use of such a great variety of programs workers have at their disposal. The process of studying the real context, abstracting the key issues in focus, and concretising them back in the form of video scenarios seemed to produce an effective result for real development. What is interesting in this case is that we did not work on any particular product, but a whole service and the overall strategy of developing knowledge management. The video-scenario practice was adopted in the bank, as they later started to produce videos themselves with an in-house team. ■

Co-relating

Whereas the previous chapters argued how video functions as designer clay and social glue in the *making* of the video artefacts, this chapter explained how the artefacts – as *presentations* – benefit from this two-fold understanding: firstly, to be effective, the video presentations need to be gripping and to show the potential to live up to people's expectations. The examples showed how the various projects drew attention to different aspects in the construction of these presentations. Secondly, the social setting around which the presentations are shown plays a crucial role in effectively influencing what people think, and hence in the facilitation of transforming the social atmosphere towards realising the visions.

A beneficial way of thinking about the role of highlight tapes and vision movies is that of *co-relating*: in the process of *relating* new observations, ideas and visions to present thinking and practices, people come to see needs and opportunities for change. In a *collaborative* viewing, such change – being inherently a social endeavour – has a better chance of succeeding.

Aftermath

Throughout this book we have taken the stance that video in design is best thought of as a malleable design material rather than as objective user data, as expressed in the metaphor of *designer clay.* The methods sections have suggested a variety of examples of how this plays out in the form of video cards, collages, portraits, scenarios and vision movies. At the same time we have shown how designers, by thinking of video as *social glue,* can employ video as a means to support design as the social process of collaboratively exploring, creating and relating in multi-disciplinary teams, with users and with other stakeholders in the design project. The case stories and video samples included on the DVD illustrate how particular conditions shape the opportunities to engage video. Employed in these ways, video indeed has the capability of focussing the user-centred design process.

Video as a technology today is developing rapidly. With video cameras becoming ubiquitously embedded in mobile phones, video editing made easy in portable equipment, and streaming on the web made available to everyone, the development of attitudes towards video use is following suit. We believe, though, that the design practices suggested in this book, and the thinking behind them, have the capacity to live on for a little longer than the market lifetime of the newest video camera models.

Designing with video   No matter the state of the technology, designing with video is a very powerful practice that needs to be learned and developed through hands-on practice: through a constant, reflective experimentation with ways of engaging people in design moves. Best of all, this endeavour is also a very enjoyable one!

References, index
and DVD contents

References

13407, I. (1999), *Human-centred design processes for interactive systems*, European standard EN ISO 13407:1999, Technical report, International Organization for Standardization (ISO).

Aarts, E. and Marzano, S. (2003), *The new everyday: Ambient intelligence in all its aspects*, 010 Publishers, the Netherlands.

Anderson, R.J. (1994), Representations and requirements: The value of ethnography in system design, *Human–Computer Interaction 9*, 151–182.

Anderson, S.; Kobara, S.; Mathis, B. and Shafrir, E. (1992), *Imagine... A Vision of Health Care in 1997*, Hewlett-Packard Company.

Archer, B. (1984 [1965]), *Developments in design methodology*, Wiley, New York, chapter Systematic method for designers, pp. 57–82.

Battarbee, K. (2004), *Co-Experience: Understanding user experiences in social interaction*, PhD thesis, University of Art and Design Helsinki.

Bergman, E.; Lund, A.; Dubberly, H.; Tognazzini, B. and Intille, S. (2004), Video visions of the future: a critical review, in *CHI 2004: CHI 2004 Extended Abstracts on Human Factors in Computing Systems*, ACM Press, New York, pp. 1584–1585.

Beyer, H. and Holtzblatt, K. (1998), *Contextual Design: A Customer-Centered Approach to Systems Designs*, Morgan Kaufmann, San Francisco, CA.

Binder, T. (1999), Setting the stage for improvised video scenarios, in *CHI 1999: CHI 1999 Extended Abstracts on Human Factors in Computing Systems*, ACM Press, New York, pp. 230–231.

Blomberg, J.; Giagomi, J.; Mosher, A. and Swenton-Wall, P. (1993), Ethnographic field methods and their relation to design, in Schuler, D. and Namioka, A. (eds.),*Participatory Design: Principles and Practices*, Lawrence Erlbaum Associates, Hillsdale, pp. 123–156.

Blumer, H. (1998 [1969]), *Symbolic Interactionism: Perspective and Method*, University of California Press.

Boal, A. (2000 [1979]), *Theatre of the Oppressed*, Pluto Press, London.

Boal, A. (1992), *Games for Actors and Nonactors*, Routledge, London.

Boal, A. (1995), *The Rainbow of Desire: The Boal Method to Theatre and Therapy*, Routledge, London.

Boal, A. (1998), *Legislative Theatre: Using Performance to Make Politics*, Routledge, London.

Brandt, E. (2006), Designing exploratory design games: A framework for participation in Participatory Design?, in *PDC 2006: Proceedings of the Ninth Conference on Participatory design*, ACM Press, New York, pp. 57–66.

Brandt, E. and Grunnet, C. (2000), Evoking the future: Drama and props in user centered design, in *Proceedings of Participatory Design Conference (PDC 2000)*, CPSR, Palo Alto, CA, pp. 11–20.

Bruner, J. (1986), *Actual Minds, Possible Worlds*, Harvard University Press, Cambridge, MA.

Buur, J.; Jensen, M.V. and Djajadiningrat, T. (2004), Hands-only scenarios and video action walls: Novel methods for tangible user interaction design, in *DIS 2004: Proceedings of the 2004 Conference on Designing Interactive Systems*, ACM Press, New York, pp. 185–192.

Buur, J. and Søndergaard, A. (2000), Video Card Game: An Augmented Environment for User Centred Design Discussions, in *Proceedings of Designing Augmented Reality Environments (DARE 2000)*, ACM Press, New York.

Bødker, S. and Iversen, O. (2002), Staging a professional participatory design practice: moving PD beyond the initial fascination of user involvement, in *Proceedings of the Second Nordic Conference on Human–Computer Interaction*, ACM Press, New York, pp. 11–18.

Cagan, J. and Vogel, C.M. (2002), *Creating Breakthrough Products: Innovation from Product Planning to Program Approval*, FT Press, Indianapolis, IN.

Collins COBUILD English Language Dictionary (1987), William Collins Sons and Co Ltd, London.

Crabtree, A. (1998), Ethnography in Participatory Design, in Chatfield, R.; Kuhn, S. and Muller, M. (eds.), *Proceedings of the 1998 Participatory Design Conference*, CPSR, Palo Alto, CA, pp. 93–105.

Crabtree, A. (2003), *Designing Collaborative Systems: A Practical Guide to Ethnography*, Springer, London.

Crabtree, A.; Hemmings, T. and Rodden, T. (2002), Pattern-based Support for Interactive Design in Domestic Settings, in *Proceedings of Designing Interactive Systems Conference DIS 2002*, ACM Press, New York, pp. 265–275.

Cross, N. (1972), Here comes everyman, in *Proceedings of the Design Participation Conference*, pp. 11–14.

Desmet, P. (2002), *Designing emotions*, PhD thesis, Delft: Delft University of Technology, the Netherlands.

Dewey, J. (1991 [1910]), *How We Think*, Prometheus Books, Amherst, NY.

Dourish, P. (2001), *Where the Action Is: The Foundations of Embodied Interaction*, MIT Press, Cambridge, MA.

Dreyfuss, H. (1967), *Designing for People*, Paragraphic Books, Grossman Publishers, New York.

Dubberly, H. and Mitch, D. (1987), *Knowledge Navigator*, Apple Computer.

Dumas, J.S. (2003), User-based evaluations, in Jacko, J.A. and Sears, A. (eds.), *The Human–Computer Interaction Handbook: Fundamentals, Evolving Technologies and Emerging Applications*, Lawrence Erlbaum Associates, Hillsdale, pp. 1094–1115.

Dumas, J.S. and Redish, J. (1993), *A Practical Guide to Usability Testing*, Ablex Publishing Co., Greenwich, CT.

Ehn, P. and Kyng, M. (1987), The collective resource approach to systems design, in Bjerknes, G., Ehn, P. and Kyng, M. (eds.), *Computers and Democracy – a Scandinavian Challenge*, Avebury, Gower Publishing Company Ltd., Aldershot, UK, pp. 17–57.

Ehn, P. and Sjögren, D. (1991), From system descriptions to scripts for action, in Greenbaum, J. and Kyng, M. (eds.), *Design at Work: Cooperative Design of Computer Systems*, Lawrence Erlbaum Associates, Hillsdale, pp. 241–268.

Ellis, J.C. (1979), *A History of Film*, Prentice-Hall, Saddle River, NJ.

Engeström, Y. (2001), *Expansive Learning at Work: Toward an activity theoretical reconceptualization*, Journal of Education and Work 14(1), 133–156.

Festinger, L. (1957), *A Theory of Cognitive Dissonance*, Stanford University Press, Stanford, CA.

Garrett, J.J. (2002), *The Elements of User Experience: User-centered Design for the Web*, New Riders Press, Indianapolis, IN.

Gaver, B.; Dunne, T. and Pacenti, E. (1999), Design: Cultural probes, in *Interactions*, Volume 6, Issue 1, ACM Press, New York, pp. 21–29.

Geertz, C. (1973), Thick description: Toward an interpretive theory of culture, in Geertz, C. (ed.), *The Interpretation of Cultures: Selected Essays*, Basic Books, New York, pp. 3–30.

Greenbaum, J. and Kyng, M. (eds.) (1991), *Design at work: Cooperative design of computer systems*, Lawrence Erlbaum Associates, Hillsdale, NJ.

Hallnäs, L. and Redström, J. (2006), *Interaction design: Foundations, experiments*, The Interactive Institute, The Swedish School of Textiles, University College of Borås. Available at: http://www.slow-technology.se/book/.

Heath, C. and Luff, P. (2000), *Technology in Action*, Cambridge University Press, UK.

Heskett, J. (2002), *Toothpicks & Logos: Design in Everyday Life*, Oxford University Press, UK.

Hughes, J.; King, V.; Rodden, T. and Andersen, H. (1994), Moving out from the control room: ethnography in system design, in *CSCW 1994: Proceedings of the 1994 ACM Conference on Computer Supported Cooperative Work*, ACM Press, New York, pp. 429–439.

Hulkko, S.; Mattelmäki, T.; Virtanen, K. and Keinonen, T. (2004), Mobile Probes, in A. Hyrsykari (ed.), *Proceedings of the Third Nordic Conference on Human–Computer Interaction*, Tampere, Finland, ACM Press, New York.

Hutchby, I. and Wooffitt, R. (1998), *Conversation Analysis: Principles, Practices and Applications*, Polity Press, Oxford.

Iacucci, G.; Kuutti, K. and Ranta, M. (2000), On the move with a magic thing: role playing in concept design of mobile services and devices, in *DIS 2000: Proceedings of the Conference on Designing Interactive Systems*, ACM Press, New York, pp. 193–202.

Isomursu, M.; Kuutti, K. and Väimämö, S. (2004), Experience clip: Method for user participation and evaluation of mobile concepts, *PDC 2004: Proceedings of the Eighth Conference on Participatory Design*, ACM Press, New York, pp. 83–92.

Ivens, J. (1969), *The Camera and I*, Seven Seas Books, Berlin, Germany.

Jacucci, G. (2004), *Interaction as performance: Cases of configuring physical interfaces in mixed media*, PhD thesis, Faculty of Science, Department of Information Processing Science, University of Oulu Acta Universitatis Ouluensis, A427.

Johnstone, K. (1987), *Impro: Improvisation and the Theatre*, Theatre Arts Books, New York.

Jordan, B. and Henderson, A. (1995), Interaction analysis: Foundations and practice, *The Journal of the Learning Sciences* 4(1), 39–103.

Jordan, P.W. (2000), *Designing Pleasurable Products: An Introduction to the New Human Factors*, Taylor & Francis, London.

Kankainen, A. (2003), UCPCD: User-centered Product Concept Design, in *DUX 2003: Proceedings of the 2003 Conference on Designing for User Experiences*, ACM Press, New York, pp. 1–13.

Keinonen, T. and Takala, R. (eds.) (2006), *Product Concept Design: A Review of the Conceptual Design of Products in Industry*, Springer, London.

Kelley, T. (2001), *The Art of Innovation: Lessons in Creativity from IDEO, America's Leading Design Firm*, Currency, New York.

Kjærdsgaard, M. and Petersen, G. (2007), *Participant Intervention and Tangible Mediating Tools*, Forthcoming.

Krippendorff, K. (1996), On the essential contexts of artifacts or on the proposition that design is making sense (of things), in Margolin, V. and Buchanan, R. (eds.), *The Idea of Design, A Design Issues Reader*, The MIT Press, Cambridge, pp. 156–184.

Kurvinen, E. (2007), *Prototyping social action*, PhD thesis, School of Design, University of Art and Design Helsinki.

Laurel, B. (1993), *Computers as Theatre*, Addison-Wesley Longman Publishing Co., Inc., Boston, MA.

Lewin, K. (1947) Group Decisions and Social Change, in Newcomb, T. and Hartley, E. (eds.), *Readings in Social Psychology*, Holt, Rinehart, and Winston.

Leyda, J. (ed.) (1970), *The Film Sense: Sergei Eisenstein*, Faber and Faber, London.

236

Lindholm, C. and Keinonen, T. (2003), Managing the design of user interfaces, in Lindholm, C., Keinonen, T. and Kiljander, H. (eds.), *How Nokia Changed the Face of the Mobile Phone*, McGraw-Hill, New York, pp. 139–154.

Macdonald, K. and Cousins, M. (1996), *Imagining Reality: The Faber Book of Documentary*, Faber and Faber, London.

Mackay, W.; Ratzer, A.V. and Janecek, P. (2000), Video artifacts for design: Bridging the gap between abstraction and detail, in *Conference on Designing Interactive Systems DIS 2000*, ACM Press, New York, pp. 72–82.

Mackay, W.E. and Fayard, A.L. (1999), Video brainstorming and prototyping: Techniques for participatory design, in *CHI 1999: CHI 1999 Extended Abstracts on Human Factors in Computing Systems*, ACM Press, New York, pp. 118–119.

Martin, D. and Sommerville, I. (2004), Patterns of cooperative interaction: Linking ethnomethodology and design, *ACM Transactions on Computer–Human Interaction*, vol. 11, no. 1, ACM Press, New York, pp. 59–89.

Mattelmäki, T. (2006), *Design probes*, PhD thesis, University of Art and Design Helsinki, Publication series A69.

May, R. (1975), *The Courage to Create*, Bantam Books, New York.

Millen, D.R. (2000), Rapid ethnography: Time deepening strategies for HCI field research, *DIS 2000: Proceedings of the Conference on Designing Interactive Systems*, ACM Press, New York, 280–286.

Muller, M. (1991), *Proceedings of ACM Conference on Human Factors in Computing Systems*, ACM Press, New Orleans, LA, chapter PICTIVE: An exploration in participatory design, pp. 225–231.

Muller, M. (1992), Retrospective on a year of participatory design using the PICTIVE technique in *Proceedings of the SIGCHI Conference on Human Factors in Computing Systems*, ACM Press, New York, pp. 455–462.

Nielsen, J. (1993), *Usability Engineering*, Academic Press, Boston, MA.

Nielsen, J. and Mack, R.L. (eds.) (1994), *Usability Inspection Methods*, Wiley, New York.

Norman, D. (1988), *The Design of Everyday Things, Basic Books*, New York.

Pine, B.J. and Gilmore, J.H. (1999), *The Experience Economy: Work is Theatre and Every Business a Stage*, Harvard Business School Press, Cambridge, MA.

Pink, S. (2001), *Doing Visual Ethnography*, Sage Publications, London.

Raijmakers, B.; Gaver, W.W. and Bishay, J. (2006), Design documentaries: Inspiring design research through documentary film, in *DIS 2006: Proceedings of the 6th ACM Conference on Designing Interactive Systems*, ACM Press, New York, pp. 229–238.

Rhea, D.K. (1992), A new perspective on design: Focusing on customer experience, *Design Management Journal* Fall, 40–48.

Rittel, H. and Webber, M. (1984), Planning problems are wicked problems, in Cross, N. (ed.), *Developments in Design Methodology*, Wiley, New York, pp. 135–144.

Rosenthal, A. (1996), *Writing, Directing, and Producing Documentary Films and Videos*, Southern Illinois University Press.

Ryle, G. (1968), The Thinking of Thoughts: What Is "Le Penseur" Doing?, *University of Saskatchewan University Lectures*, no. 18. (25.1.2007) Available at: http://lucy.ukc.ac.uk/CSACSIA/Vol14/Papers/ryle_1.html.

Sanders, E.B.-N. (2001), Virtuosos of the experience domain, in *Proceedings of the 2001 IDSA Education Conference*. (11.6.2007) Available at http://www.maketools.com/.

Sanders, E.B.-N. and Dandavate, U. (1999), Designing for experiencing: New tools, in Overbeeke, C.J. and Hekkert, P. (eds.), *Proceedings of the First International Conference on Design and Emotion*, TU Delft.

Schuler, D. and Namioka, A. (eds.) (1993), *Participatory Design: Principles and Practices*, Lawrence Erlbaum Associates, Hillsdale, NJ.

Schön, D. (1983), *The Reflective Practitioner: How Professionals Think in Action*, Basic Books, New York.

Schön, D. (1987), *Educating the Reflective Practitioner*, Jossey-Bass, San Francisco, CA.

Shrum, W.; Duque, R. and Brown, T. (2005), Digital video as research practice: Methodology for the millennium, *Online Journal of Research Practice* 1(1), Article M4. Available at: http://www.icaap.org/.

Sperschneider, W. (2000),*Filmportrait Fredrik Barth: From Fieldwork to Theory*, Institute for Scientific Film, Video, 60 min., Göttingen.

Sperschneider, W. and Bagger, K. (2000), Ethnographic fieldwork under industrial constraints: Towards design-in-context, in *The Proceedings of NordiCHI 2000*.

Stienstra, M. (2003), *Is every kid having fun? A gender approach to interactive toy design*, PhD thesis, Centre for Studies of Science, Technology and Society, Twente University.

Suchman, L.A. (1987), *Plans and Situated Actions: The Problem of Human–Machine Communication*, Cambridge University Press, New York.

Suchman, L.A. and Trigg, R. (1991), *Design at Work: Cooperative Design of Computer Systems, Lawrence Erlbaum Associates*, chapter Understanding practice: Video as a medium for reflection and design, pp. 65–90.

Svanæs, D. and Seland, G. (2004), Putting the users center stage: Role playing and low-fi prototyping enable end users to design mobile systems, in *CHI 2004: Proceedings of the SIGCHI Conference on Human Factors in Computing Systems*, ACM Press, New York, pp. 479–486.

Tognazzini, B. (1994), The "Starfire" video prototype project: A case history, in *Proceedings of ACM Conference on Human Factors in Computing Systems*, ACM Press, Boston, MA, pp. 99–105.

Ulrich, K.T. and Eppinger, S.D. (2003), *Product Design and Development*, Mc-Graw-Hill, New York.

Vertelney, L. (1989), Using video to prototype user interfaces, *SIGCHI Bulletin* October, 57–61.

Visser, F.S.; Stappers, J.; van der Lugt, R. and Sanders, E.B.N. (2005), Context-mapping: Experiences from practice, *CoDesign: International Journal of CoCreation in Design and the Arts* 1(2), 119–149.

Wenger, E. (1998), *Communities of Practice: Learning, Meaning and Identity*, Cambridge University Press, New York.

Whalen, M.; Whalen, J.; Moore, R.J.; Raymond, G.T.; Szymanski, M.H. and Vinkhuyzen, E. (2004), Studying workscapes, in Levine, P. and Scollon, R. (eds.), *Discourse & Technology: Multimodal Discourse Analysis*, Georgetown University Press, Washington, DC, pp. 208–229.

Winston, B. (1995), *Claiming the Real: The Griersonian Documentary and Its Legitimations*, British Film Institute.

Wittgenstein, L. (1976), *Philosophical Investigations*, Basil Blackwell & Mott, Ltd., Oxford, UK.

Ylirisku, S. and Vaajakallio, K. (2007), Situated make tools for envisioning ICTs with ageing workers, in *Proceedings of Include 2007: Designing with People Conference*.

Index

DVD contents

The video clips on the enclosed DVD are original excerpts from real design projects. As they include materials from VHS tapes, digital clips, mobile phone clips, etc., their visual and audio quality varies substantially.

Printed in the United States
By Bookmasters